MONOGRAPHS ON
PHYSICAL BIOCHEMISTRY

GENERAL EDITORS

W. F. HARRINGTON A. R. PEACOCKE

MIGRATION OF INTERACTING SYSTEMS

BY

L. W. NICHOL

Professor of Physical Biochemistry
Australian National University

AND

D. J. WINZOR

Reader in Biochemistry
University of Queensland

CLARENDON PRESS · OXFORD

1972

Oxford University Press, Ely House, London W.1

GLASGOW NEW YORK TORONTO MELBOURNE WELLINGTON
CAPE TOWN IBADAN NAIROBI DAR ES SALAAM LUSAKA ADDIS ABABA
DELHI BOMBAY CALCUTTA MADRAS KARACHI LAHORE DACCA
KUALA LUMPUR SINGAPORE HONG KONG TOKYO

PRINTED IN GREAT BRITAIN
BY BUTLER AND TANNER LTD, FROME AND LONDON

PREFACE

THE binding of a relatively small molecule or ion, termed the 'ligand' to a relatively large molecule, the 'acceptor', is of wide occurrence in biochemical systems. It is not surprising, therefore, that considerable attention has been and is being given to the characterization of such systems. At first sight the problem is formidable since many acceptor and ligand systems are known to self-interact by a variety of reactions including isomerizations, polymerizations, hybridizations, and heterogeneous associations. At the same time the problem is of interest because these interactions may well form the basis of a variety of biological phenomena, such as the sigmoidal binding and kinetic curves important in certain aspects of metabolic control.

In the early descriptions of the way in which interactions would be manifested in familiar experimental procedures, such as the mass migration processes of electrophoresis, sedimentation velocity, and chromatography, warnings were issued against facile interpretation of results in terms of classical theories established to handle non-interacting systems. In some quarters the response was to select an environmental medium which inhibited the interaction, and while valuable results emerged pertaining, for example, to the molecular weights of protein subunits, it has become increasingly clear that the interactions themselves are of considerable biological importance and should therefore be elucidated.

In recent years experimental procedures and theories have been extended to permit a positive approach to the investigation of such systems: the equilibrium methods of light scattering, osmotic pressure, sedimentation equilibrium, and equilibrium dialysis may be employed to gain information on thermodynamic parameters, while other procedures permit comment on the rates of reaction and on the chemical nature of binding sites and intermolecular bond types. These important aspects receive only brief mention in this text, which limits itself to a discussion of the utility of mass migration experiments in characterizing interacting systems. In presenting methods for analysing results (usually with the aim of evaluating thermodynamic parameters), we have stressed the theoretical background first and followed it in each section with a practical example to illustrate the point being made. The choice of these examples was not arbitrary; but rather was guided by the interests of the authors in protein research. We believe, however, that the suggested procedures could be employed with advantage in the study of other types of system involving polysaccharides, nucleic acids, or synthetic polymers and their interactions with a variety of ligands ranging from hormones to acridine dyes. In this connection an understanding of the principles of the theoretical background seems more important than a detailed knowledge of the reactions of a particular class of compounds.

Moreover, in our opinion, the presentation would be sterile if it conveyed the impression that the evaluation of thermodynamic parameters was the sole aim of studying interacting systems, intrinsically interesting and challenging though this might be. The first chapter therefore discusses the forms and curve-fitting of binding curves found with a variety of acceptor–ligand systems: it becomes apparent that the binding equations describing these curves involve the thermodynamic parameters, and the need for their evaluation is thereby established. The mass migration methods developed in succeeding chapters will be seen to be capable not only of providing binding results, but also of evaluating the parameters required for their interpretation.

ACKNOWLEDGEMENTS

PART of this manuscript was prepared while L.W.N. was a member of the Russell Grimwade School of Biochemistry, University of Melbourne.

We are indebted to the following publishers for permitting the reproduction of copyright material: American Chemical Society Publications, Academic Press, Elsevier Publishing Company, Macmillan (Journals) Ltd., the Biochemical Society, and the Royal Society.

CONTENTS

1

MACROMOLECULAR INTERACTIONS AND BIOLOGICAL FUNCTION

THE field of macromolecular chemistry provides a multitude of examples of systems which in solution comprise an equilibrium mixture of different forms of the macromolecular entity (for example, Reithel 1963; Nichol, Bethune, Kegeles, and Hess 1964). The different types of equilibrium interaction encountered may be summarized as follows:

$n\mathrm{A} \rightleftharpoons \mathrm{C}$ isomerization ($n=1$), or polymerization ($n>1$);

$n\mathrm{A}+m\mathrm{B} \rightleftharpoons \mathrm{A}_n\,\mathrm{B}_m$ complexing of dissimilar molecules;

$$\left.\begin{array}{l}\mathrm{A}+\mathrm{B} \rightleftharpoons \mathrm{C}+\mathrm{D}\\ p_1p_2+q_1q_2 \rightleftharpoons p_1q_1+p_2q_2\end{array}\right\}\ \text{hybridization.}$$

These cited reactions may be extended or combined to describe more complicated situations that may arise. For example, weight-average molecular weight studies in particular environments show that monomeric forms of α-chymotrypsin (Rao and Kegeles 1958) and glutamate dehydrogenase (Frieden and Colman 1967, Eisenberg and Tomkins 1968) coexist in equilibrium with a *series* of higher polymeric forms, and accordingly a single equilibrium constant is insufficient to describe the system. Nevertheless, despite its simplicity, the above classification of reaction types permits a systematic investigation of the behaviour of interacting systems when subjected to mass migration under the influence of an applied potential gradient, as in electrophoresis, velocity sedimentation, or chromatography. In cases where the rate of re-equilibration is rapid in comparison with the differential rates of migration of the individual species, the interpretation of the resulting migration pattern is undeniably complicated. An instance is the schlieren pattern observed in velocity sedimentation of a rapidly polymerizing protein system; regardless of whether the pattern exhibits a single peak or partial resolution, sedimentation coefficients measured from maximum ordinate positions cannot be used directly to assess hydrodynamic parameters of individual polymeric forms, because the observed boundary represents concentration gradients of *all* polymeric forms. However, in favourable circumstances the experimental pattern may be interpreted in terms not only of velocities pertaining to individual species, but also of thermodynamic parameters characterizing the association equilibrium. Such characterization of an interacting system is a worth-while endeavour in itself; but it assumes greater significance when it is noted that combinations of interactions of the types cited may provide explanations of several biologically important phenomena, such as sigmoidal binding or activity curves found with proteins implicated in the control of metabolic pathways. The sigmoidal binding curves of oxygen to haemoglobin (for example, Wyman 1964) or to the haemocyanins of certain invertebrates (Moore, Henderson, and Nichol 1968) are examples of the importance of the sigmoidal response in respiratory control: other biological implications of similar sigmoidal effects have been reviewed by Umbarger (1964), Atkinson (1965), Stadtman (1966), and Koshland and Neet (1968). We shall discuss in detail some bases of regulation in order to illustrate more clearly the types of interacting system involved, and to establish at the outset the value of migration studies on these systems.

1.1. Combination of equilibrium reactions

Consider first a system in which a relatively large *acceptor* molecule isomerizes or polymerizes in solution according to $n\mathrm{A} \rightleftharpoons \mathrm{C}$. To the solution is added a relatively small ligand which binds to p sites per molecule of A and to q sites per molecule of C: the multiple binding specified involves formation of complexes between dissimilar molecules and is denoted by $\leftrightarrow$. Since many ligand molecules, such as organic dyes, N-acetyl-D-glucosamine, adenosine-5′-phosphate, cholesterol, and chlorophylls, are known to either isomerize or polymerize (reviewed, for example, in Nichol, Smith, and Ogston 1969*b*), it is realistic also to consider a ligand self-interaction, written as $m\mathrm{S} \rightleftharpoons \mathrm{T}$. The complete model is summarized as follows:

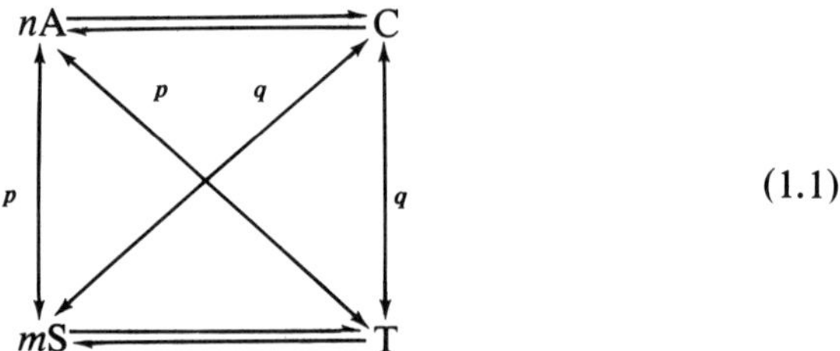

(1.1)

At equilibrium there exists in the solution unbound ligand, S and T, and a series of complexes of both A and C: those of the former are denoted by $\mathrm{AS}_i\mathrm{T}_j$, where $0 \leqq i+j \leqq p$ ($i+j=0$ represents unreacted A), a similar notation being adopted with C ($\mathrm{CS}_k\mathrm{T}_l$, $0 \leqq k+l \leqq q$, $k+l=0$ represents unreacted C).

The investigator may determine the *total* weight concentrations of acceptor and ligand (in whatever form) in the mixture, $\bar{c}_P$ and $\bar{c}_L$, respectively: these are best expressed on the g per unit volume scale, because the only unambiguous alternative when polymerizing systems are encountered is the less direct base-molar scale. Provided the concentration of unbound (or free) ligand summed over all forms, $\bar{c}$ (g per unit volume), may also be determined, it is possible to evaluate the corresponding value of the binding function, r, defined as the number of grammes of ligand bound per gramme of acceptor, i.e.

$$r = \frac{(\bar{c}_L - \bar{c})}{\bar{c}_P}. \tag{1.2}$$

A series of experiments (preferably performed, as we shall see, holding $\bar{c}_P$ fixed and varying $\bar{c}_L$) gives a set of $(r, \bar{c})$ results, which may be plotted directly to give a *binding curve.* The experimental determination of $\bar{c}$ is obviously basic to the construction of such a curve, and as an alternative to the familiar method of equilibrium dialysis it will be shown in Chapter 2 that a carefully designed migration method of frontal analysis in gel chromatography may be employed.

1.2. Binding equations

The ultimate aim of a binding study is to curve-fit the experimental binding curve on the basis of a *binding equation* (expressing the functional relationship between r and $\bar{c}$ in terms of physical parameters pertinent to the system) derived from a scheme such as eqn (1.1).

(a) *A particular example: intrinsic binding constant*

For the purpose of illustration, suppose in relation to eqn (1.1) that $p = q = 2$, and accordingly that the acceptor entities may be written as —A— and —C— to denote binding sites available for either S or T. There will coexist in the equilibrium mixture unbound —A—, —C—, S, and T, together with the series of complexes S—A—, —A—S, T—A—, —A—T, S—A—T, T—A—S, S—A—S, T—A—T, and a similar set involving the acceptor C. If it is assumed that all sites within each acceptor entity are equivalent and independent with regard to the binding of a particular ligand species, we may write

$$\begin{aligned}
m_{\text{S-A-}} &= m_{\text{-A-S}} \doteq K_A m_A m_S, \\
m_{\text{T-A-}} &= m_{\text{-A-T}} = L_A m_A m_T, \\
m_{\text{S-A-T}} &= m_{\text{T-A-S}} = K_A L_A m_A m_S m_T, \\
m_{\text{S-A-S}} &= K_A^2 m_A m_S^2, \\
m_{\text{T-A-T}} &= L_A^2 m_A m_T^2.
\end{aligned} \tag{1.3}$$

The first equation of this set specifies the *equivalence* of sites on A in that the same equilibrium constant (K_A) describes the binding of S to either site. The concept of *independent* (i.e. not mutually interacting) sites is introduced in the fourth equation of the set, where K_A is again employed to describe the second addition of ligand, regardless of the position occupied by the first ligand molecule. Equation (1.3) defines the intrinsic binding constants K_A and L_A (litres/mol, with m denoting molar concentrations) (Klotz 1946, 1953), and a similar set of equations may be written to define K_C and L_C, the intrinsic binding constants relating to the binding of S and T, respectively, to C. In writing intrinsic binding constants, as with all other equilibrium constants to be defined later, it is assumed that activity coefficients of all entities are unity. The basic information required to formulate the required binding equation is summarized in Table 1.1. Although binding sites are assumed to be equivalent, they are distinct physically, and thus, as illustrated in eqn (1.3), the various stoichiometric complexes are actually constituted by chemically distinct entities, the number of which is given by the statistical factor. It could also be noted that the equilibrium concentration of any stoichiometric complex is independent of the pathway of formation (for example, the equilibrium concentration of the complex AST is not affected by the order in which the ligand molecules bind). With the use of the information in Table 1.1 it is possible to write expressions for the total concentrations (g/litre) of acceptor and ligand, and to note the simplifications (by factorization) made possible by the introduction of the concepts of equivalent independent sites and related intrinsic binding constants.

TABLE 1.1

Expressions for formulation of the binding equation describing a system involving two binding sites for S *and* T *on* A *and* C

Stoichiometric complex	Statistical factor	Moles of constituent per mole of complex		Equilibrium concentration of each complex species
		A	S	(M)
AS	2	1	1	$K_A m_A m_S$
AT	2	1	m	$L_A m_A m_T$
AST	2	1	$(1+m)$	$K_A L_A m_A m_S m_T$
AS_2	1	1	2	$K_A^2 m_A m_S^2$
AT_2	1	1	$2m$	$L_A^2 m_A m_T^2$
CS	2	n	1	$K_C m_C m_S$
CT	2	n	m	$L_C m_C m_T$
CST	2	n	$(1+m)$	$K_C L_C m_C m_S m_T$
CS_2	1	n	2	$K_C^2 m_C m_S^2$
CT_2	1	n	$2m$	$L_C^2 m_C m_T^2$

$$\bar{c}_P = M_A(m_A + 2K_A m_A m_S + 2L_A m_A m_T + 2K_A L_A m_A m_S m_T + K_A^2 m_A m_S^2 + L_A^2 m_A m_T^2) + nM_A(m_C + 2K_C m_C m_S + 2L_C m_C m_T + 2K_C L_C m_C m_S m_T + K_C^2 m_C m_S^2 + L_C^2 m_C m_T^2),$$

$$\bar{c}_P = M_A m_A(1 + K_A m_S + L_A m_T)^2 + nM_A m_C(1 + K_C m_S + L_C m_T)^2. \quad (1.4)$$

$$\bar{c}_L = M_S\{m_S + mm_T + 2K_A m_A m_S + 2mL_A m_A m_T + 2(1+m)K_A L_A m_A m_S m_T + 2K_A^2 m_A m_S^2 + 2mL_A^2 m_A m_T^2 + 2K_C m_C m_S + 2mL_C m_C m_T + 2(1+m)K_C L_C m_C m_S m_T + 2K_C^2 m_C m_S^2 + 2mL_C^2 m_C m_T^2\},$$

$$\bar{c}_L = M_S(m_S + mm_T) + 2M_S m_A(1 + K_A m_S + L_A m_T)(K_A m_S + mL_A m_T) + 2M_S m_C(1 + K_C m_S + L_C m_T)(K_C m_S + mL_C m_T) \quad (1.5)$$

Noting that M_A and M_S are the molecular weights of monomeric forms of acceptor and ligand, respectively, and that $\bar{c} = M_S(m_S + mm_T)$, we may combine eqns (1.2), (1.4), and (1.5) to give the final binding equation

$$r = \frac{2M_S m_A(1 + K_A m_S + L_A m_T)(K_A m_S + mL_A m_T) + 2M_S m_C(1 + K_C m_S + L_C m_T)(K_C m_S + mL_C m_T)}{M_A m_A(1 + K_A m_S + L_A m_T)^2 + nM_A m_C(1 + K_C m_S + L_C m_T)^2} \quad (1.6)$$

(b) *General formulation*

The above procedure may be generalized for the case, as in eqn (1.1), where there are p equivalent binding sites per molecule of A and q equivalent sites per molecule of C: the procedure extends that given by Nichol *et al.* (1969*b*).

$$\bar{c}_P = M_A\left\{\sum_{i+j=0}^{i+j=p} m_A(K_A m_S)^i(L_A m_T)^j w_{i,j} + n\sum_{k+l=0}^{k+l=q} m_C(K_C m_S)^k(L_C m_T)^l w_{k,l}\right\}. \quad (1.7)$$

The parameters $w_{i,j}$ and $w_{k,l}$ are the statistical weights of the complexes of composition $AS_i T_j$ and $CS_k T_l$, respectively, and are equivalent to statistical factors: they are given by

$$w_{i,j} = \frac{p!}{i!j!(p-i-j)!}, \quad (1.8a)$$

$$w_{k,l} = \frac{q!}{k!l!(q-k-l)!}. \quad (1.8b)$$

It is clear that the values of $w_{i,j}$ and $w_{k,l}$ are coefficients of a binomial expansion, and thus eqn (1.7) may be simplified to

$$\bar{c}_P = M_A m_A(1 + K_A m_S + L_A m_T)^p + nM_A m_C(1 + K_C m_S + L_C m_T)^q. \quad (1.9)$$

Similar expressions may be written for the weight-concentration of bound ligand as follows:

$$\begin{aligned}
\bar{c}_{\mathrm{L}}-M_{\mathrm{S}}(m_{\mathrm{S}}+mm_{\mathrm{T}})
&=M_{\mathrm{S}}\{\sum(i+mj)m_{\mathrm{AS}_i\mathrm{T}_j}w_{i,j}+\sum(k+ml)m_{\mathrm{CS}_k\mathrm{T}_l}w_{k,l}\}\\
&=pM_{\mathrm{S}}m_{\mathrm{A}}\left\{K_{\mathrm{A}}m_{\mathrm{S}}\sum(K_{\mathrm{A}}m_{\mathrm{S}})^{i-1}(L_{\mathrm{A}}m_{\mathrm{T}})^{j}\frac{(p-1)!}{(i-1)!j!(p-i-j)!}\right.\\
&\quad\left.+mL_{\mathrm{A}}m_{\mathrm{T}}\sum(K_{\mathrm{A}}m_{\mathrm{S}})^{i}(L_{\mathrm{A}}m_{\mathrm{T}})^{j-1}\frac{(p-1)!}{i!(j-1)!(p-i-j)!}\right\}\\
&\quad+qM_{\mathrm{S}}m_{\mathrm{C}}\left\{K_{\mathrm{C}}m_{\mathrm{S}}\sum(K_{\mathrm{C}}m_{\mathrm{S}})^{k-1}(L_{\mathrm{C}}m_{\mathrm{T}})^{l}\frac{(q-1)!}{(k-1)!l!(q-k-l)!}\right.\\
&\quad\left.+mL_{\mathrm{C}}m_{\mathrm{T}}\sum(K_{\mathrm{C}}m_{\mathrm{S}})^{k}(L_{\mathrm{C}}m_{\mathrm{T}})^{l-1}\frac{(q-1)!}{k!(l-1)!(q-k-l)!}\right\}\\
&=pM_{\mathrm{S}}m_{\mathrm{A}}(K_{\mathrm{A}}m_{\mathrm{S}}+mL_{\mathrm{A}}m_{\mathrm{T}})(1+K_{\mathrm{A}}m_{\mathrm{S}}+L_{\mathrm{A}}m_{\mathrm{T}})^{p-1}\\
&\quad+qM_{\mathrm{S}}m_{\mathrm{C}}(K_{\mathrm{C}}m_{\mathrm{S}}+mL_{\mathrm{C}}m_{\mathrm{T}})(1+K_{\mathrm{C}}m_{\mathrm{S}}+L_{\mathrm{C}}m_{\mathrm{T}})^{q-1}.
\end{aligned}\tag{1.10}$$

The summations in eqn (1.10) are from $(i+j)=1$ to $(i+j)=p$ and from $(k+l)=1$ to $(k+l)=q$: thus no terms for unbound A or C are included. Combination of eqns (1.2), (1.9), and (1.10) yields the more general binding equation

$$r=\frac{pM_{\mathrm{S}}m_{\mathrm{A}}(K_{\mathrm{A}}m_{\mathrm{S}}+mL_{\mathrm{A}}m_{\mathrm{T}})(1+K_{\mathrm{A}}m_{\mathrm{S}}+L_{\mathrm{A}}m_{\mathrm{T}})^{p-1}+qM_{\mathrm{S}}m_{\mathrm{C}}(K_{\mathrm{C}}m_{\mathrm{S}}+mL_{\mathrm{C}}m_{\mathrm{T}})(1+K_{\mathrm{C}}m_{\mathrm{S}}+L_{\mathrm{C}}m_{\mathrm{T}})^{q-1}}{M_{\mathrm{A}}m_{\mathrm{A}}(1+K_{\mathrm{A}}m_{\mathrm{S}}+L_{\mathrm{A}}m_{\mathrm{T}})^{p}+nM_{\mathrm{A}}m_{\mathrm{C}}(1+K_{\mathrm{C}}m_{\mathrm{S}}+L_{\mathrm{C}}m_{\mathrm{T}})^{q}}\tag{1.11}$$

Evidently eqn (1.6) is consistent with eqn (1.11) for the special case that $p=q=2$.

Finally, it is possible to generalize further the result to encompass a situation in which a *series* of ligand species, i, each bind to a *series* of acceptor molecules, j. If β_j is the number of binding sites on the jth acceptor species $(j=1, 2, \ldots y)$ and M_j is its molecular weight, with m_j the molar equilibrium concentration of uncomplexed jth acceptor species and $\tau_{i,j}$ the intrinsic binding constant of the ith ligand to the jth acceptor, we may write

$$r=\frac{\sum_{j=1}^{j=y}\left\{\beta_j m_j\left(1+\sum_{i=1}^{i=x}\tau_{i,j}m_i\right)^{\beta_j-1}\sum_{i=1}^{i=x}M_i\tau_{i,j}m_i\right\}}{\sum_{j=1}^{j=y}\left\{M_j m_j\left(1+\sum_{i=1}^{i=x}\tau_{i,j}m_i\right)^{\beta_j}\right\}}.\tag{1.12}$$

(c) *Simplifications of binding equations and numerical examples*

(i) *The rectangular hyperbola.* By placing m_C and m_T equal to zero in eqn (1.11) the physical situation described is the binding of a non-selfinteracting ligand to p equivalent and independent sites on a non-selfinteracting acceptor entity (i.e. $S \leftrightarrow A$). The plot of r vs. m_S (or $\bar{c} = M_S m_S$) is a rectangular hyperbola described by the equation

$$r = \frac{M_S p K_A m_S}{M_A(1 + K_A m_S)}. \tag{1.13}$$

For this simple case a *double-reciprocal plot* of binding results ($1/r$ vs. $1/m_S$) gives a straight line, the slope ($M_A/pM_S K_A$) and intercept (M_A/pM_S) of which enable p and K_A to be estimated provided M_A and M_S are known. Alternatively, a plot of r/m_S vs. r, termed a *Scatchard plot* (Scatchard 1949), yields according to eqn (1.13) a straight line of slope $-K_A$ and abscissa intercept $(r=0)M_S pK_A$. The experimental observation of linear double-reciprocal or Scatchard plots does not guarantee, however, that the simple model $S \leftrightarrow A$ applies.

Consider a situation in which $m_C = 0$ and $m = 1$, i.e. the ligand isomerizes with a dimensionless isomerization constant $Y = m_T/m_S$, and both isomers bind to p equivalent sites on a non-selfinteracting acceptor entity. Equation (1.11) becomes, on substituting $m_S = \bar{c}/M_S(1+Y)$,

$$r = \frac{pM_S \bar{c}(K_A + YL_A)/(1+Y)}{M_A\{M_S + \bar{c}(K_A + YL_A)/(1+Y)\}}. \tag{1.14}$$

A plot of r vs. $\bar{c}$ (the total weight concentration of unbound ligand), which would be constructed in practice, is still of the form of a rectangular hyperbola, and the corresponding double-reciprocal plot would thus be linear of slope $M_A(1+Y)/p(K_A + YL_A)$ and ordinate intercept M_A/pM_S. Isomerization of the ligand cannot therefore be detected from the *form* of a binding curve, though its occurrence will affect the significance of the constants that describe such curves. In this situation, the experimenter may only determine p and K_A if he has evaluated Y from studies of the ligand system alone. This consideration could be of importance, for example, in the interpretation of apparent intrinsic binding constants obtained for the binding of N-acetyl-D-glucosamine to lysozyme (Adkins and Yang 1968), since the mutarotation of the ligand was not considered. It has been shown that α and β anomers of certain derivatives of this ligand possess different binding affinities for lysozyme (Raftery, Dahlquist, Chan, and Parsons 1968).

If we place $m_T = 0$, $K_A = K_C$ and $q = np$ in eqn (1.11), we now encounter a system in which a non-selfinteracting ligand multiply binds to equal extents to two isomeric or polymeric forms of the acceptor. Equation (1.11) then simplifies directly to eqn (1.13); in this case the values of K_A ($=K_C$) and p ($=q/n$) evaluated from the linear double-reciprocal plot suffice to describe the system in terms of binding parameters, provided the value of n is established from separate studies on the self-interacting acceptor system alone. Finally, in a special case noted first by Wyman (1967), when $m_T=0$, $n=1$, and $p=q=1$, eqn (1.11) simplifies to

$$r=\frac{M_S(K_A+XK_C)m_S}{M_A(1+X)+M_A m_S(K_A+XK_C)}, \tag{1.15}$$

where $X=m_C/m_A$ is the isomerization constant of acceptor. Thus, in this special case of one binding site on acceptor *isomers* a binding curve with the form of a rectangular hyperbola is obtained regardless of whether $K_A=K_C$ or $K_A \neq K_C$. Again the system may only be characterized completely by studying the acceptor isomerization to determine X.

(ii) *Sigmoidal curves.* Let us continue to examine situations where $m_T=0$ so that we have a non-selfinteracting ligand binding to an acceptor which is isomerizing or polymerizing. Except in the circumstances noted above, eqn (1.11) cannot be simplified to a form describing a rectangular hyperbola. However, recourse may be made to numerical examples to illustrate the form of a binding curve described by the complicated binding equation. It is clear that values assigned to relevant parameters should not be such as to make $K_A=K_C$ and $q=np$, for then a simple rectangular hyperbola would result. Deviations from this simple form will occur when the ligand binds *preferentially* to one form of the acceptor. It is convenient (though not essential) to make C a non-binding form by placing *either* q or K_C equal to zero. A thermodynamic treatment does not require a microscopic interpretation of the latter relations; but it may be helpful to suggest that when the acceptor isomerizes, sites are preserved ($q=p$) but distorted in one isomeric form such that $K_A \neq K_C$, while acceptor polymerization may directly involve binding sites such that $q \neq np$. Regardless of these tentative explanations, it follows that when $m_T=0$ and q and/or $K_C=0$, eqn (1.11) becomes

$$r=\frac{pM_S m_A K_A m_S(1+K_A m_S)^{p-1}}{M_A m_A(1+K_A m_S)^p+nM_A m_C}. \tag{1.16}$$

The procedure for construction of a numerical example involves assigning values to p, M_S, K_A, M_A, and n, and calculating from eqn (1.16) values of r corresponding with a suitable range of m_S values. A problem arises in that the molar equilibrium concentrations of unbound forms of acceptor, m_A and m_C, cannot be assigned arbitrary values since they are functionally related to m_S

by the law of mass action. When $n=1$ the problem is simply solved by dividing throughout by m_A and assigning a value to $X=m_C/m_A$, which now appears in the second term of the denominator. This was the basic procedure employed by Monod, Wyman, and Changeux (1965) in their construction of sigmoidal binding curves for the isomerizing acceptor case (shown in their Fig. 1(a), (b)). It is noteworthy that the same binding equation derived on a thermodynamic basis applies to the Monod model and that stipulating a pre-existing polymerization of the acceptor prior to the addition of ligand. Moreover, the examples of Monod et al. (1965), based on the concept of intrinsic binding constants and hence on *equivalence and independence* of binding sites within an acceptor entity, show that certain sigmoidal effects may be explained, at least macroscopically, without invoking cooperative or mutually interacting sites. However, with other systems the possibility is not denied that binding of ligand at one site may *induce* conformational changes which perturb the structure of the acceptor so that the affinity for ligand at other sites is altered (for example, Koshland, Némethy, and Filmer 1966; Koshland and Neet 1968).

When $n>1$ (acceptor polymerization prior to the addition of ligand) the problem relating to the construction of numerical examples remains because, since $X=m_C/m_A^n$, division of eqn (1.16) by m_A does not eliminate both m_A and m_C as variables. Fortunately, a second simultaneous equation is available on noting that the denominator of eqn (1.16) has been defined as $\bar{c}_P$, the total acceptor concentration in the equilibrium mixture, and may be written as

$$nM_A X m_A^n + M_A m_A (1+K_A m_S)^p - \bar{c}_P = 0. \qquad (1.17)$$

For an assigned value of $\bar{c}_P$ the polynomial may be solved for m_A for a range of m_S values, and the corresponding values of m_C ($m_C = X m_A^n$) and r (from eqn (1.16)) evaluated. When $n=2$ (monomer–dimer system) eqn (1.17) is a quadratic in m_A, only the positive root being acceptable; thus

$$m_A = \frac{-M_A(1+K_A m_S)^p + \{M_A^2(1+K_A m_S)^{2p} + 4nM_A X\bar{c}_P\}^{1/2}}{2nM_A X}. \qquad (1.18)$$

Placing $\alpha = K_A m_S$ and $m_C/m_A = Xm_A$ in eqn (1.16) gives eqn (1.19), which describes ligand binding to a monomer–dimer system in which dimer is inactive ($K_C=0$):

$$r = \frac{pM_S\alpha(1+\alpha)^{p-1}}{M_A(1+\alpha)^p + nM_A X m_A}. \qquad (1.19)$$

The transform $\alpha = K_A m_S$ is useful because selection of α as abscissa of a binding curve permits the representation of several cases with different K_A

values by a single curve. Moreover, inspection of the simultaneous eqns (1.18) and (1.19) shows X and $\bar{c}_P$ appear in the final solution only as the product $X\bar{c}_P$, and thus calculated examples for different values of this product illustrate the effect of either a variation in X at constant $\bar{c}_P$ or a variation in $\bar{c}_P$ at constant X. Examples of binding curves calculated on the basis of eqns (1.18) and (1.19) for different values of $X\bar{c}_P$ are shown in Fig. 1.1. Two points in relation to Fig. 1.1 merit comment. First it is clear that a model based on polymerization of acceptor and preferential ligand binding provides (as does the isomerization model of Monod) a plausible basis for sigmoidal binding curves. Second,

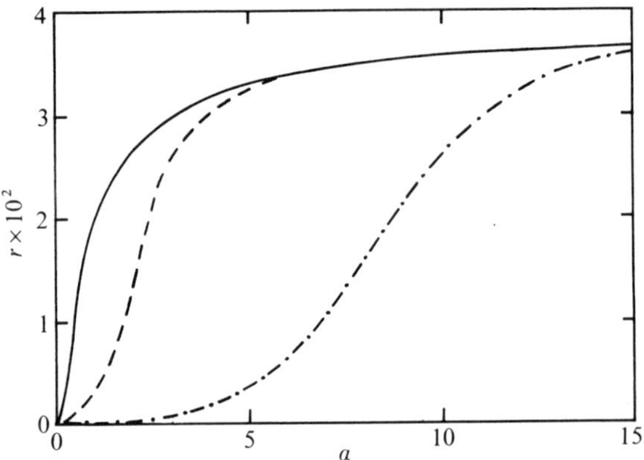

FIG. 1.1. Binding curves for a system in which a non-self-interacting ligand of molecular weight 200 ($M_S = 200$) binds to four equivalent and independent sites ($p = 4$) on an acceptor monomer ($M_A = 20\ 000$) which coexists with a non-binding dimer ($n = 2$, q and/or $K_C = 0$). Values of the product $X\bar{c}_P$, made dimensionless by expressing X in litres/g, are as follows: ———, 1; – – – –, 10^4; – · – · – · –, 10^8. (Adapted from Nichol et al. (1967*a*).)

comparison of the curves shows that the extent of sigmoidality increases as the value of $X\bar{c}_P$ increases (i.e. $dr/d\alpha$ increases less rapidly at low α the larger the value of $X\bar{c}_P$ but all curves reach the same maximum limiting value of r). This finding may be viewed in two ways. Comparison at fixed $\bar{c}_P$ of systems characterized by different association equilibrium constants, X, illustrates that pronounced sigmoidality is observed with those systems with a large excess of non-binding dimer in equilibrium with active monomer prior to the addition of ligand. It follows that evolutionary changes of the amino-acid sequence of certain proteins may not be directed solely towards the formation of a structure possessing an efficient active site: at least for some proteins involved in metabolic control a structure conducive to the formation of a large reservoir of inactive polymeric (or isomeric) form would be advantageous. In this connection, numerical examples calculated for different values of n (see Fig. 2 of Nichol, Jackson, and Winzor 1967*a*) suggest that polymerization equilibria, especially those characterized by a large value of n, are more effective than isomerizations in producing sigmoidality. However, it is hazardous to ascribe merit to a particular form of control on the basis of numerical examples: suffice it to stress that self-interaction of acceptor (for example, isomerization or polymerization), together with preferential binding of ligand to that form present initially in smaller relative amount, produces sigmoidal effects.

Re-examination of Fig. 1.1 from the viewpoint that a particular system (characterized by a fixed value of X) is being studied at different total acceptor concentrations, $\bar{c}_P$, illustrates that the forms of binding curves relating to a polymerizing system change with $\bar{c}_P$. The effect in this particular example is, of course, exaggerated, since the range of acceptor concentrations required to effect the variation of $X\bar{c}_P$ could not be realized experimentally. Nevertheless, concentration-dependence of binding results is characteristic of a polymerizing acceptor system, a feature which distinguishes it both theoretically (via eqns (1.18) and (1.19) and experimentally from an isomerizing system, which is described solely by eqn (1.16). In this regard there is very little basis for the claim that dissociation (polymerization) reactions of proteins may be considered as a special case of subunit interactions (Nichol 1965, Koshland and Neet 1968). The result also reinforces the suggestion made earlier that binding (or enzymic activity) curves should be constructed by varying $\bar{c}_L$ at fixed $\bar{c}_P$. For example, if a kineticist, while attempting to establish a Michaelis–Menten curve for the enzyme system under consideration, were to increase $\bar{c}_P$ as well as varying substrate concentration, the result may well be a curve exhibiting a maximum, erroneously suggesting excess substrate inhibition: in fact, the two-dimensional plot obtained would be a traverse of a family of sigmoidal curves of the type illustrated in Fig. 1.1.

Finally, it is important to note that the present treatment permits construction of numerical examples in cases where the polymeric (rather than the monomeric) form of acceptor exclusively binds ligand. In this case, increasing sigmoidality is observed with decreasing $\bar{c}_P$ (at fixed X), since dilution favours the non-binding form. It also permits consideration of examples where both (or several) forms of the acceptor bind ligand to different extents. For the case of a non-self-interacting ligand ($m_T = 0$) and two forms of acceptor the relevant expressions are

$$r = \frac{pM_S K_A m_A m_S(1+K_A m_S)^{p-1} + qM_S K_C m_C m_S(1+K_C m_S)^{q-1}}{M_A m_A(1+K_A m_S)^p + nM_A m_C(1+K_C m_S)^q}, \quad (1.20a)$$

$$X = \frac{m_{\mathrm{C}}}{m_{\mathrm{A}}^{n}}, \tag{1.20b}$$

$$nXM_{\mathrm{A}}(1 + K_{\mathrm{C}}m_{\mathrm{S}})^{q}m_{\mathrm{A}}^{n} + M_{\mathrm{A}}(1 + K_{\mathrm{A}}m_{\mathrm{S}})^{p}m_{\mathrm{A}} - \bar{c}_{\mathrm{P}} = 0. \tag{1.20c}$$

Their solution requires assigning values to q and K_{C} in addition to those for the special case of exclusive binding (Fig. 1.1). In the event that a series of acceptor entities are encountered, a series of values of β_j and $\tau_{i,j}$ must be assigned in accordance with eqn (1.12).

(iii) *The effect of self-interaction of ligand.* Equations (1.11) or (1.12) may be simplified in a variety of ways to permit construction of numerical examples illustrating the effects of self-interaction of ligand. For example, if the polymeric form of a ligand (and not the monomeric species) binds to a non-self-interacting acceptor, we place m_{C} and K_{A} equal to zero in eqn (1.11), and obtain

$$r = \frac{pM_{\mathrm{S}}mL_{\mathrm{A}}m_{\mathrm{T}}}{M_{\mathrm{A}}(1 + L_{\mathrm{A}}m_{\mathrm{T}})} \tag{1.21}$$

for the model $m\mathrm{S} \rightleftharpoons \mathrm{T} \leftrightarrow \mathrm{A}$. It is clear from eqn (1.21) that a plot of r vs. m_{T} would be a rectangular hyperbola; but this is not usually relevant, since an experimenter can only determine the *total* concentration of unbound ligand, $\bar{c}$, given by

$$\bar{c} = M_{\mathrm{S}}(m_{\mathrm{S}} + mm_{\mathrm{T}}) = M_{\mathrm{S}}\left\{\left(\frac{m_{\mathrm{T}}}{Y}\right)^{1/m} + mm_{\mathrm{T}}\right\}. \tag{1.22}$$

For a range of values of m_{T}, eqn (1.21) yields r and eqn (1.22) the corresponding values of $\bar{c}$. Except when $m = 1$ (and eqn (1.14) applies), calculated curves of r vs. $\bar{c}$ are sigmoidal (see Fig. 1(b) of Nichol *et al.* 1969*b*). As these workers noted, however, exclusive binding of the *monomeric* form of ligand to a non-self-interacting acceptor generates an r–$\bar{c}$ plot which deviates in form from a rectangular hyperbola, but not in the direction of sigmoidality. In addition, they found that self-interaction of ligand (other than isomerization) may operate to accentuate or obscure a sigmoidal effect inherent in the multiple binding of ligand to a *self-interacting acceptor.*

Let us consider one example of the latter type, described by

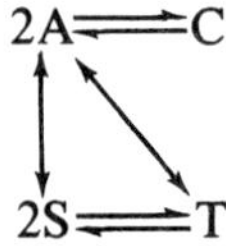

The relevant binding equation is obtained readily from eqn (1.11) by substituting $K_C = L_C = 0$ and/or $q = 0$, the required simultaneous equation being formed by equating the denominator of eqn (1.11) to $\bar{c}_P$. The results of numerical calculations are shown in Fig. 1.2, where values assigned to relevant para-

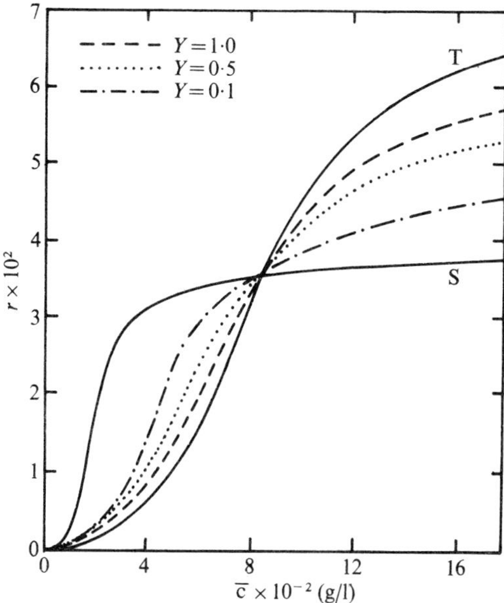

FIG. 1.2. The effect of varying the equilibrium constant, Y, governing the dimerization of ligand in a system where both monomer and dimer bind with identical intrinsic association constants to monomeric acceptor, but neither binds to the dimeric form of acceptor. The following values of parameters apply to all curves: $M_A = 20\,000$, $M_S = 200$, $p = 4$, $n = 2$, $X = 10^7\,M^{-1}$, $\bar{c}_P = 10$ g/litre, $m = 2$, $K_A = L_A = 1\,M^{-1}$, $K_C = L_C = 0$ and/or $q = 0$. The solid lines correspond to situations where ligand exists in only one form (either as monomer or as dimer). (Adapted from Nichol *et al.* (1969*b*).)

meters are also reported. The sigmoidal curves (———) obtained when $Y = 0$ (only S is present) and $Y = \infty$ (only T is present) are expected in these situations from eqns (1.16) to (1.19) inclusive. The different maximum limiting values for r for the solid curves arise simply because the amount (g) of dimer bound to four sites on A at saturation is greater than the corresponding amount of bound monomer. It is apparent that when both ligand forms bind to A with $K_A = L_A$ that sigmoidal curves persist, and that small changes in the equilibrium constant Y are manifested as measurable changes in the sigmoidal binding curves, which range between the extremes set by the solid curves. Figure 1.1 has already established that any factor (such as pH variation or the action of drugs, hormones, or feedback inhibitors) which affect the value of X, pertaining to acceptor self-interaction, will alter the form of the sigmoidal curve. It is now clear from Fig. 1.2, that similar factors affecting the value of Y, pertaining to polymerization of ligand, also offer a fine means of adjustment of the forms of sigmoidal curves.

1.3. Curve-fitting sigmoidal binding results

While it is instructive to employ derived binding equations for the construction of exact numerical examples, it is more useful to utilize the equations to elucidate the basis of sigmoidal binding data obtained experimentally. With the precision of binding results presently attainable, it is unlikely that any curve-fitting procedure could define a unique set of binding parameters when the number of such parameters is large, as with certain situations encompassed by eqn (1.12). In this section we shall illustrate the approaches to curve-fitting which have been adopted with simpler systems.

(a) *Isomerizing acceptor systems*

We visualize a situation in which an investigator has obtained a sigmoidal binding curve and established from separate experiments on acceptor and ligand alone that the acceptor isomerizes but that the ligand does not self-interact. The fact that a *sigmoidal* binding curve independent of $\bar{c}_P$ has been obtained immediately excludes the possibility that $p=q=1$ or $p=q>1$ with $K_A=K_C$. It also shows that the preferential binding of ligand to one isomeric form is competitive with acceptor isomerization inasmuch as addition of ligand causes a shift in equilibrium towards the more active isomer and its complexes. The latter point may be seen more clearly by writing from the denominator of eqn (1.20a) expressions for the molar concentrations of unbound isomer A and all of the complexes involving A, $\bar{m}_A$, and a similar expression for the corresponding constituent concentration $\bar{m}_C$.

$$\bar{m}_A=m_A(1+K_A m_S)^p, \tag{1.23a}$$

$$\bar{m}_C=m_C(1+K_C m_S)^q. \tag{1.23b}$$

It follows that

$$\frac{\bar{m}_C}{\bar{m}_A^n}=\frac{m_C(1+K_C m_S)^q}{m_A^n(1+K_A m_S)^{np}}. \tag{1.24}$$

Thus the ratio $\bar{m}_C/\bar{m}_A$ (with $n=1$) remains constant (and equal to $X=m_C/m_A$) only when $K_A=K_C$ and $q=np$, in which case, as we have already seen, the binding curve assumes the form of a rectangular hyperbola. In other cases with $q\neq np$ and/or $K_A\neq K_C$, the ratio $\bar{m}_C/\bar{m}_A$ changes with m_S and a sigmoidal curve results. Since p and q are merely labels in the case of an isomerizing system, specification of the more active isomer is not required: for convenience, let it be A. At extremely high ligand concentrations the ratio $\bar{m}_C/\bar{m}_A$ approaches zero, and from eqn (1.20a),

$$\lim_{m_S \to \infty} r = \frac{pM_S}{M_A}, \tag{1.25}$$

which is the maximum limiting value of r obtained with a simple rectangular hyperbola (eqn (1.13)). The first step in curve-fitting sigmoidal binding results obtained with an isomerizing acceptor system is to plot $1/r$ vs. $1/m_S$ and draw the limiting tangent to the curve (which appears *concave* to the $1/m_S$ axis) as $1/m_S \to 0$. From the slope $(M_A/pM_S K_A)$ and intercept (M_A/pM_S) of the tangent, values of p and K_A may be evaluated. This procedure is illustrated in Fig. 1.3(a) with binding data obtained by Changeux, Gerhart, and Schachman (1968) for the system aspartate transcarbamylase-succinate (a competitive inhibitor) in the presence of 4 mM carbamyl phosphate. Values of M_S and M_A have already been incorporated into the binding data, which are expressed in terms of a binding function defined on a molar scale $(r_M = rM_A/M_S)$. In this instance, values of 3·7 for p and 2·0 mM^{-1} for K_A are found, which compare favourably with the corresponding values reported by Changeux and Rubin (1968), viz. 3·8 and 2·0 mM^{-1}, respectively, which were obtained by an equivalent extrapolation of a Scatchard plot of the same data.

The remaining unknowns in eqns (1.20a) and (1.20b) are q, K_C, and X. In principle, it may appear possible to substitute experimental values for three sets of (r, m_S) data into the equation and solve the resulting simultaneous equations for these unknowns. However, q appears as an exponent in these expressions, and it is therefore simplest to assign values to this parameter, whereupon corresponding values of K_C and X may be calculated from different combinations of (r, m_S) data. Unacceptable values of q may then be detected by lack of agreement between estimates of apparent equilibrium constants so obtained, or, in some instances, by failure of the two simultaneous equations to yield solutions that are meaningful physically. For the aspartate transcarbamylase-succinate system the only acceptable solution appears to be q and/or $K_C = 0$ and $X = 4$ (Changeux and Rubin 1968; Nichol, Smith, and Winzor 1969*c*).

Although the aspartate transcarbamylase system was introduced here merely to illustrate a general approach to curve-fitting, it is relevant to comment on some general features of the result. First, we note the recent reports (Weber 1968, Wiley and Lipscomb 1968) of evidence for six regulatory and six catalytic subunits in the enzyme, whereupon consideration of symmetry factors would lead to an expected value of 6 for p: from the

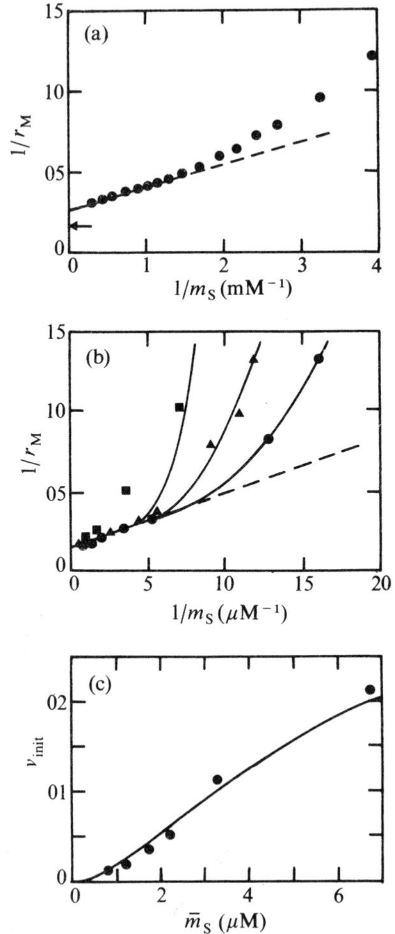

FIG. 1.3. Illustrations of curve-fitting procedures. (a) Double-reciprocal plot of sigmoidal binding data obtained with the aspartate transcarbamylase-succinate system (adapted from Nichol *et al.* (1969*c*).) (b) Double reciprocal plots of sigmoidal binding data for the glutamate dehydrogenase-GTP system. Solid lines have been calculated on the basis of a monomer–hexamer system with $p=6$, $K_A=5\cdot0\ \mu M^{-1}$, $q=0$, $X=6\cdot7\times10^{36}\ M^{-5}$, and the constituent concentrations of enzyme relevant to the three sets of experimental results: ●, 0·12 mg/ml; ▲, 0·30 mg/ml; ■, 1·0 mg/ml. (Adapted from Nichol *et al.* (1969*c*).) (c) The sigmoidal kinetics of firefly luciferase. Circles represent experimental values, while the curve refers to a system with the following characteristics: $p=1$, $L_A=10^5\ M^{-1}$, $\bar{m}_B=2\cdot5\,\bar{m}_S$, $K=2\cdot5\times10^5\ M^{-1}$, $V_{max}=0\cdot6$, and $\bar{m}_A=1\cdot07\times10^{-7}$ M. (Adapted from Smith and Nichol (1971).)

theoretical ordinate intercept of the double-reciprocal plot for a system with $p=6$ (arrow in Fig. 1.3(a)) it is clear that the experimental binding data do not conform with such a model. Second, the indicated value of $X=4$, specifying in accordance with previous findings an excess of nonbinding isomer C, does not necessarily pertain to an equilibrium between isomeric forms of the enzyme in the *absence* of carbamyl phosphate. The presence of a second substrate introduces complications which will be discussed later. Third, there appears to be no *direct* evidence that, even in the presence of carbamyl phosphate, the enzyme exists as an equilibrium mixture of isomeric forms prior to the addition of the inhibitor, succinate, or substrate (aspartate). The decrease of 3·5 per cent in the weight-average sedimentation coefficient obtained under the conditions of the binding experiments in the presence and absence of succinate (Gerhart and Schachman 1968) is in accord *either* with a shift in the ratio $\bar{m}_C/\bar{m}_A$ as applies to binding that is competitive with a pre-existing isomerization, *or* with a ligand-induced conformational change as proposed by Koshland (1958, 1960, 1963, 1964) and favoured for this system by Markus, McClintock and Bussell (1971). Moreover, a distinction between the possibilities cannot be made on the basis of weight-average molecular weight studies, which would exhibit a lack of concentration-dependence regardless of whether an isomerization equilibrium existed or whether the enzyme (prior to addition of ligand) behaved as a non-interacting solute. No such uncertainty with regard to the basic model and parameters derived therefrom need necessarily apply when polymerization equilibria are the bases of observed sigmoidal binding effects.

(b) *Polymerizing acceptor systems*

The procedure used to curve-fit binding results obtained with a system in which acceptor polymerizes and ligand does not self-interact is similar to that described above, and will be illustrated with data pertaining to the binding of guanosine triphosphate (GTP) to glutamate dehydrogenase (GDH) in the presence of 4×10^{-4} M NADH (Frieden and Colman 1967). First, it is noted that the weight-average molecular weight of the enzyme is concentration-dependent in a way compatible with polymerization of the enzyme. In addition, the weight-average molecular weight decreases on addition of GTP, showing in accord with eqn (1.24) that GDH monomer binds GTP preferentially. Accordingly, from eqn (1.25) it follows that the tangent to the binding data, plotted in double-reciprocal form, as $1/m_S \to 0$ provides values of p and K_A. Fig. 1.3(b) presents the binding data reported for three concentrations of GDH, values of r_M having been adjusted to conform with a value of 310 000 for the molecular weight of the enzyme (Eisenberg and Tomkins 1968) rather than the value of 400 000 employed originally by Frieden and Colman (1967). From the slope and intercept of the limiting tangent (broken line in Fig. 1.3(b)) values of 6 and $5{\cdot}0\ \mu M^{-1}$ are estimated for p and K_A, respectively.

The nature of the polymeric oligomer of GDH is not known for certain, but for the purpose of illustration the binding data have been interpreted (Nichol *et al.* 1969*c*) on the basis of an inactive hexamer, since molecular weights in the vicinity of 1 800 000 to 2 000 000 have been obtained at high concentrations of the enzyme. The equations relevant to this first attempt at curve-fitting are eqns (1.16) and (1.17), which with known values of p, K_A, n, M_A, and M_S (=440), may be solved for values of m_A and X corresponding to a particular experimental point (r_M, m_S) read from a binding curve obtained with a known value of $\bar{c}_P$. The procedure was used (Nichol *et al.* 1969*c*) to obtain an apparent value of $6{\cdot}7 \times 10^{36}\ M^{-5}$ from the binding data obtained with $\bar{c}_P = 0{\cdot}12$ mg/ml (circles in Fig. 1.3(b)). This value was then employed in eqns (1.16) and (1.17) to compute a theoretical double-reciprocal plot, which is shown as the solid line in Fig. 1.3(b) and is seen to describe the experimental data very well over the entire range of substrate concentrations. These estimates of q (=0) and X ($=6{\cdot}7 \times 10^{36}\ M^{-5}$) also provide a reasonable description of the experimental data at the second lowest enzyme concentration (denoted by triangles), but much poorer agreement between prediction and experiment is observed when an acceptor concentration of 1 mg/ml (data denoted by squares) is considered: description in terms of these parameters is completely inadequate at higher concentrations of GDH.

Inability of the same set of parameters to fit the binding data at all concentrations of acceptor could be ascribed to effects such as the possession of some affinity for GTP by the hexamer, or to the presence of other polymers besides monomer and hexamer. It would, of course, be possible to write a set of simultaneous equations based on eqn (1.12) to account for a series of polymers binding GTP to different extents, but their solution would be extremely difficult. Moreover, there is another possible explanation of the observed discrepancy at high $\bar{c}_P$ values. Binding data were obtained in the presence of a second ligand, NADH, which is known to alter the weight-average molecular weight of GDH at a given protein concentration (Colman and Frieden 1966), and may therefore act as a modifier (or effector) of the system. Further discussion of this point follows later.

(c) *Self-interacting ligand systems*

It was noted in section 2c(iii) of this Chapter that self-interaction of ligand may lead to sigmoidal binding curves even in cases where the acceptor does not self-interact. Firefly luciferase, which displays sigmoidal kinetics in one of its reactions (Denburg and De Luca 1968) appears to be one such acceptor. With this system the substrate, a complex between a magnesium and a pyrophosphate ion, exists in equilibrium with its constituent ionic species, neither of which alone functions as a substrate for the enzyme. This system could be represented by the simple model

$$\begin{array}{c} \mathrm{A} \\ {\scriptstyle p}\updownarrow{\scriptstyle L_{\mathrm{A}}} \\ \mathrm{S}+\mathrm{B}\rightleftharpoons\mathrm{T} \end{array}$$

where the substrate complex T binds to p equivalent sites on the enzyme (A) with an intrinsic binding constant L_A.

The relevant binding equation for this system is eqn (1.21) with $m=1$ and expressions for the total molar concentrations of A, S, and B given by

$$\bar{m}_A = m_A(1+L_A m_T)^p, \tag{1.26a}$$

$$\bar{m}_S = m_S + m_T + pL_A m_A m_T(1+L_A m_T)^{p-1}, \tag{1.26b}$$

$$\bar{m}_B = m_B + m_T + pL_A m_A m_T(1+L_A m_T)^{p-1}. \tag{1.26c}$$

In addition, an association equilibrium constant describing substrate complex formation may be defined as

$$K = \frac{m_T}{m_S m_B}. \tag{1.26d}$$

The kinetic data of Denburg and De Luca (1968) have been curve-fitted (Smith and Nichol 1971) by assuming that the initial velocity of reaction, v_{init}, is related to r_M by the expression

$$pv_{init} = V_{max} r_M, \tag{1.27}$$

where V_{max} is the maximum initial velocity of the enzymic reaction. This relationship is not without criticism, Frieden (1967) having discussed in detail the assumptions inherent therein.

Denburg and De Luca (1968) showed that with excess magnesium (B) essentially all pyrophosphate (S) was converted to the substrate complex (T), in which case the enzyme exhibited normal Michaelis–Menten kinetics. This permits the direct determination of $p=1$ and $L_A = 1\times10^5\,\mathrm{M}^{-1}$ from the double-reciprocal plot (Fig. 3 of Denburg and De Luca 1968). A value of $V_{max}=0{\cdot}6$ units is indicated by the results shown in Figs. 1 and 2 of these workers. Their quoted value of $2{\cdot}5\times10^5\,\mathrm{M}^{-1}$ for K may also be used in the calculation, together with the value $\bar{m}_A = 1{\cdot}07\times10^{-7}\,\mathrm{M}$ (Denburg and De Luca, personal communication). These values of parameters have been used to calculate v_{init} as a function of the total pyrophosphate concentration, $\bar{m}_S$, employing eqns (1.21), (1.26), and (1.27). The result is shown as a solid line in Fig. 1.3(c), where the experimental points of Denburg and De Luca (1968) are also included for comparison: clearly, the fit is reasonable.

1.4. Systems involving two or more ligands: modifiers

It is well-established that an entity with structure entirely different from that of substrate(s) may affect the form of a binding (or activity) curve obtained with a particular acceptor–ligand system. An example is provided by the feed-back inhibition by cytidine triphosphate (CTP) of the aspartate transcarbamylase-aspartate-carbamyl phosphate system (Gerhart and Pardee 1962, 1963). In the presence of CTP the kinetic behaviour exhibits increased sigmoidality, and the modifier (or effector), CTP, is termed an inhibitor. In contrast, the addition of adenosine triphosphate (ATP) results in decreased sigmoidality of the curve so that it approaches the form of a rectangular hyperbola, and this particular modifier (ATP) is termed an activator.

One approach to this problem is to consider the following possible model,

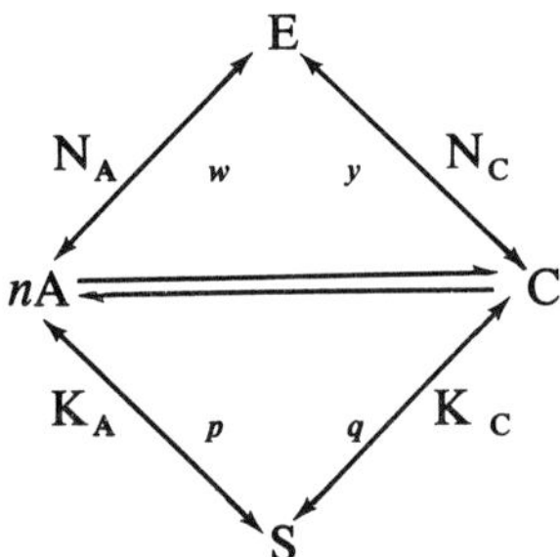

It differs from previous models involving a self-interacting ligand in that the modifier, E, is assumed to bind to w equivalent sites on A and y equivalent sites on C, which differ from the p and q sites potentially available for ligand (S) binding. In the previously considered cases involving an isomerizing or polymerizing ligand it was reasonably assumed that the various oligomeric forms of the ligand competed for the same sites, if they bound to acceptor at all. To stress this difference, let us take a particular example (chosen quite arbitrarily) in which $p=w=2$ and $q=y=1$, so that we may represent sites available to S by solid lines and those available to E by broken lines, and write

In an equilibrium mixture, as well as unbound A, C, S, and E, there will coexist a series of complexes, which are indicated in Table 1.2, together with statistical factors and other relationships (as in Table 1.1) useful for the formulation of the binding equation. As before, it is a simple matter to sum the appropriate

TABLE 1.2

Expressions for the formulation of the *binding equation describing a* system *involving the binding of a ligand and a modifier to a self-interacting acceptor* $(p=w=2,\ q=y=1)$

Stoichiometric complex	Statistical factor	Moles of constituent per mole of complex A	S	E	Equilibrium concentration of each complex species
AS	2	1	1	0	$K_A m_A m_S$
AS_2	1	1	2	0	$K_A^2 m_A m_S^2$
AE	2	1	0	1	$N_A m_A m_E$
AE_2	1	1	0	2	$N_A^2 m_A m_E^2$
ASE	4	1	1	1	$K_A N_A m_A m_S m_E$
AS_2E	2	1	2	1	$K_A^2 N_A m_A m_E m_S^2$
ASE_2	2	1	1	2	$K_A N_A^2 m_A m_S m_E^2$
AS_2E_2	1	1	2	2	$K_A^2 N_A^2 m_A m_S^2 m_E^2$
CS	1	n	1	0	$K_C m_C m_S$
CE	1	n	0	1	$N_C m_C m_E$
CSE	1	n	1	1	$K_C N_C m_C m_S m_E$

molar equilibrium concentrations (multiplied by the corresponding statistical factors and stoichiometry coefficients) to obtain expressions for the total concentrations (g/litre) $\bar{c}_P$, $\bar{c}_L$ (or $\bar{c}_S$), and $\bar{c}_E$. Reiteration of the procedure for different values of p, q, w, and y leads, by mathematical induction, to the general expressions (Nichol, O'Dea and Baghurst 1972b),

$$\bar{c}_P = M_A\{m_A(1+K_A m_S)^p(1+N_A m_E)^w + \\ + n m_C(1+K_C m_S)^q(1+N_C m_E)^y\}, \tag{1.28a}$$

$$\bar{c}_L(=\bar{c}_S) = M_S\{m_S + pK_A m_A m_S(1+K_A m_S)^{p-1}(1+N_A m_E)^w + \\ + qK_C m_C m_S(1+K_C m_S)^{q-1}(1+N_C m_E)^y\}, \tag{1.28b}$$

$$\bar{c}_E = M_E m_E\{1 + wN_A m_A(1+N_A m_E)^{w-1}(1+K_A m_S)^p + \\ + yN_C m_C(1+N_C m_E)^{y-1}(1+K_C m_S)^q\}. \tag{1.28c}$$

The reader may care to verify that the general eqns (1.28a, b, c) are consistent with the special case $p=w=2$ and $q=y=1$ (expressions given in Table 1.2). It is also a simple matter to utilize eqn (1.2) with $\bar{c}=M_S m_S$ to write a binding equation relating the amount of S bound per gramme of acceptor as a function of m_S. The binding equation so obtained contains expressions involving m_E, the concentration of unbound modifier, whereupon it becomes apparent that in the construction of numerical examples, or in a curve-fitting procedure based on the model, the worker must employ the binding equation (together with $X=m_C/m_A^n$) and *both* simultaneous eqns (1.28a) and (1.28c). In other words, known values of $\bar{c}_P$ *and* $\bar{c}_E$ (both held constant in the experimental construction of the r vs. m_S curve) are required to characterize the system. Construction of numerical examples for fixed $\bar{c}_P$ values shows that the variation of $\bar{c}_E$ affects the binding curve in the following manner. When the model is selected to ensure that E binds preferentially to that form of the acceptor which also binds S preferentially, stepwise increase in $\bar{c}_E$ results in a family of curves approaching the form of a rectangular hyperbola: thus E acts as an activator. Conversely, inhibitory effects are observed on increasing $\bar{c}_E$ if the selection of parameters is such as to ensure that E binds preferentially to the form of acceptor with least affinity for S. In effect, the action of the modifier is to vary the equilibrium constant governing the relative amounts of acceptor forms, and accordingly the behaviour described above is essentially illustrated by Fig. 1.1. This may become clearer by examining the special case considered by Frieden (1967) and Changeux and Rubin (1968), with $n=1$ and $w=y$. Combination of eqns (1.2), (1.28a), and (1.28b) gives for this case

$$r=\frac{pM_S K_A m_S(1+K_A m_S)^{p-1}+qM_S K_C X' m_S(1+K_C m_S)^{q-1}}{M_A(1+K_A m_S)^p+X'(1+K_C m_S)^q}, \quad (1.29a)$$

where $$X'=\frac{m_C(1+N_C m_E)^y}{m_A(1+N_A m_E)^y}. \quad (1.29b)$$

Although eqn (1.29a) is equivalent to eqn (1.20a) with $n=1$, it should be stressed that X' (unlike X) is not a constant, but is a function of the free concentration of modifier ($X'=f(m_E)$), which in turn depends on $\bar{c}_P$, $\bar{c}_L$, and $\bar{c}_E$. It approximates to a constant when $\bar{c}_E$ is sufficiently large so that $N_C m_E \gg 1$ and $N_A m_E \gg 1$, whereupon eqn (1.29b) becomes

$$X' \simeq X(N_C/N_A)^y. \quad (1.30)$$

Thus a single constant X' may suffice to describe binding data obtained with a range of m_S if the concentration of modifier is nearly saturating. This appears to be the case with the succinate-aspartate transcarbamylase binding data considered previously, which was obtained in the presence of 4 mM carbamyl phosphate, this concentration of modifier having been chosen to ensure saturation of the enzyme with respect to carbamyl phosphate. It must be stressed, however, that the value of 4 for the isomerization constant that is obtained by the curve-fitting procedure is a value of X' rather than of X. Similarly, the polymerization constant found for the GTP–GDH system (Fig. 1.3(b)) is also a value of X', since binding studies were performed in the presence of 0·4 mM NADH, which could act as modifier. Under conditions of fixed, but non-saturating $\bar{m}_E$, variation of X' with protein concentration would be expected, although a single value would apply over a limited range of acceptor concentration such that m_E remained approximately constant. This may well explain the observation (Nichol *et al.* 1969*c*) that divergence between experimental and theoretical curves (calculated with a single X' value) increases with increasing $\bar{c}_P$. However, a complete curve-fitting procedure has not been attempted with this system, which is further complicated by the uncertainty as to the exact nature of GDH polymerization.

An experimenter working with a system such as aspartate transcarbamylase-succinate (or aspartate) -carbamyl phosphate-CTP (or ATP) encounters a situation where more than two different types of ligand are involved. For an isomerizing acceptor system, eqn (1.29a) applies, the product in eqn (1.29b) being extended as follows. If there are i modifier species and the ith species binds to A at w equivalent sites with an intrinsic binding constant $\tau_{i,A}$, and to C at y equivalent sites with intrinsic binding constant $\tau_{i,C}$, then

$$X' = X \prod_i \left\{ \frac{(1+\tau_{i,C}m_i)^y}{(1+\tau_{i,A}m_i)^w} \right\}. \tag{1.31}$$

If in addition any of the modifiers self-interact, then the summations inside parentheses must be extended as indicated in eqn (1.12). It is thus possible to write a binding equation and the required simultaneous equations even for a complicated system involving a series of self-interacting acceptor, ligand, and modifier entities: but it would be exceedingly difficult to curve-fit binding data in terms of the numerous parameters involved in very complicated cases. Moreover, in certain circumstances it may be difficult to provide an explanation in molecular terms of the means by which ligand and effector, binding at different sites, may both operate to effectively alter the extent of the acceptor self-interaction, which has been implied in the above discussion. In this connection, the proposed model based on simple equilibrium concepts may have to be modified to include the proposals of Koshland.

1.5. Non-equivalent binding sites

It was assumed implicitly in the derivation of the general binding expression (eqn (1.11)) that binding sites within a particular acceptor entity were independent and equivalent (except for a statistical effect) with regard to binding of ligand. The treatment of systems in which binding sites are *non-equivalent* and independent may be illustrated with the simple situation in which a single, non-self-interacting acceptor, —A· ·, possesses two such sites capable of binding a non-self-interacting ligand, S. The expressions appropriate to the formulation of the binding equation are

$$m_{\text{S—A}\cdot\,\cdot}=K_1 m_A m_S,$$

$$m_{\text{—A}\cdot\,\cdot\text{S}}=K_2 m_A m_S, \tag{1.32a}$$

$$m_{\text{S—A}\cdot\,\cdot\text{S}}=K_1 K_2 m_A m_S^2,$$

$$\bar{c}_A=M_A m_A(1+K_1 m_S)(1+K_2 m_S), \tag{1.32b}$$

$$\bar{c}_S-c_S=M_S m_A(K_1 m_S+K_2 m_S+2K_1 K_2 m_S^2), \tag{1.32c}$$

where K_1 and K_2 are equilibrium constants ($K_1 \neq K_2$) and $\bar{c}_A$ is the total weight-concentration of acceptor. Combining eqns (1.32b), (1.32c), and (1.2), and utilizing the method of partial fractions, we find that

$$r=\frac{M_S}{M_A}\left\{\frac{K_1 m_S}{(1+K_1 m_S)}+\frac{K_2 m_S}{(1+K_2 m_S)}\right\}. \tag{1.33}$$

The treatment may readily be extended to a situation in which the acceptor possesses F classes of binding sites, each class being constituted of p_J equivalent sites associated with an intrinsic binding constant K_J. In this case Blake and Peacocke (1968) showed that

$$\frac{M_A r}{M_S}=\sum_{J=1}^{J=F}\frac{p_J K_J m_S}{1+K_J m_S}. \tag{1.34}$$

The latter authors discuss in detail the analysis of the curved Scatchard plot (r/m_S vs. r) to obtain values of p_J and K_J, and stress the difficulties encountered in instances where there is a continuous range of binding sites with many K_J. It might also be noted that the double-reciprocal plot of the binding results ($1/r$ vs. $1/m_S$) for the case under consideration would be convex to the $1/m_S$ axis.

Behaviour similar to that described above may also be observed in other situations. For example, when a non-self-interacting ligand binds to sites on two non-interacting solutes, A and C, in a mixture, eqn (1.20a) applies on noting the following: (i) the p sites on A are equivalent and independent, a similar restriction applying to the q sites on C; (ii) nM_A must be replaced by M_C, as no integral relation exists necessarily between the molecular weights of the different acceptor species; (iii) the mass action relationship (eqn (1.20b)) no longer applies. However, eqns (1.23a) and (1.23b) continue to apply, and may be combined with eqn (1.20a) to give

$$r=\frac{M_{\mathrm{S}}\left\{\dfrac{pK_{\mathrm{A}}m_{\mathrm{S}}\bar{m}_{\mathrm{A}}}{1+K_{\mathrm{A}}m_{\mathrm{S}}}+\dfrac{qK_{\mathrm{C}}m_{\mathrm{S}}\bar{m}_{\mathrm{C}}}{1+K_{\mathrm{C}}m_{\mathrm{S}}}\right\}}{M_{\mathrm{A}}\bar{m}_{\mathrm{A}}+M_{\mathrm{C}}\bar{m}_{\mathrm{C}}}. \tag{1.35}$$

Since $\bar{m}_{\mathrm{A}}$ and $\bar{m}_{\mathrm{C}}$, the molar constituent concentrations of the individual acceptor entities, are held constant in the series of binding experiments (which is comparable with keeping $\bar{c}_{\mathrm{P}}$ fixed), the denominator of eqn (1.35) is constant, whereupon the formal similarity between eqns (1.35) and (1.33) as the sum of two rectangular hyperbolae becomes obvious. Nichol et al. (1967*a*) noted that a double-reciprocal plot convex to the $1/m_{\mathrm{S}}$ axis could also be obtained in the binding of a non-self-interacting ligand to the mixture of polymeric or isomeric forms of the acceptor, provided the binding was preferential to one form but without the ratio $\bar{m}_{\mathrm{C}}/\bar{m}_{\mathrm{A}}^{n}$ being a function of m_{S}. This situation could arise in the event that, as a result of slowly attained equilibrium between the various forms of acceptor, detailed balance among them was not maintained during the course and measurement of binding (Weber 1965, Weber and Anderson 1965). In these cases, termed non-competitive by Nichol *et al.* (1967*a*), concentration-dependence of binding curves would be observed for a polymerizing (but not an isomerizing) acceptor system, provided sufficient time were allowed for equilibrium to be established between acceptor entities prior to the addition of ligand. An example of the type of double-reciprocal plots which might be obtained is shown in Fig. 1.4, which serves to illustrate the form of a curve convex to the abscissa and which highlights the clear distinction between such curves and those associated with sigmoidal effects (cf. Fig. 1.3(a), (b)).

1.6. Discussion

Several methods are available for investigating the binding of a ligand to an acceptor, which may be a protein or, indeed, any macromolecular solute. Some methods depend on spectral (for example, Hammes and Schimmel 1965; Hummel, Ver Ploeg, and Nelson 1961), optical rotatory (Adkins and Yang 1968), or fluorescent quenching (Velick, Parker, and Eisen 1960) perturbations inherent in the formation of acceptor–ligand complexes, and accordingly may measure specific rather than total binding. Equilibrium dialysis (for example, Klotz, Walker, and Pivan 1946; Changeux *et al.* 1968) provides a method of estimating $\bar{c}$, the total equilibrium concentration of all forms of unbound ligand, free from complications due to the joint presence of unbound acceptor and of complexes: the assay of the ligand concentration in the

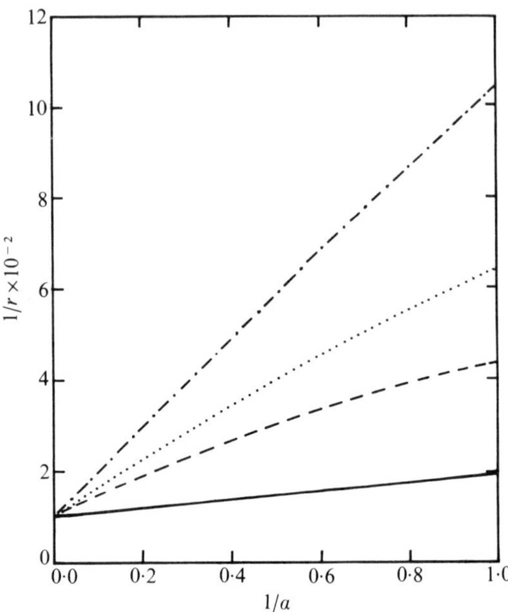

FIG. 1.4. Double-reciprocal plots of binding curves for a system in which acceptor monomer ($M_A = 20\,000$) coexists in equilibrium with dimer prior to the addition of ligand, but dimerization and ligand binding are non-competitive. Ligand ($M_S = 200$) binds to one site on monomer and two sites on dimer, with intrinsic binding constants K_A and K_C, respectively. The solid line refers to cases where $K_A = K_C$, the broken lines to $K_C/K_A = 0{\cdot}1$: – – – –, $X\bar{c}_P = 6 \times 10^8$; · · · · · ·, $X\bar{c}_P = 6 \times 10^9$; – · – · – · –, $X\bar{c}_P = 10^{12}$ (X expressed in litres/g). (Adapted from Nichol *et al.* (1967*a*).)

diffusate may utilize any spectral, optical rotatory, radio-active, or refractometric property of the ligand system. It suffers from the disadvantages that corrections must be applied for any membrane adsorption or Gibbs–Donnan effects, and that times required to attain equilibrium may be excessive in investigations of labile materials. Nevertheless, provided a membrane may be found or modified (Craig and Konigsberg 1961, Myer and Schellman 1962) to be permeable only to ligand, the method yields corresponding values of r and $\bar{c}$, according to eqn (1.2), with prior knowledge of $\bar{c}_L$ and $\bar{c}_P$. A recent application of Diaflo membranes (Blatt, Robinson, and Bixler 1968) to study binding certainly shortens the time required for each experiment, but requires $\bar{c}_L$ to be estimated in a mixture of acceptor, ligand, and complex at the completion of the experiment. Other methods employed to evaluate $(r, \bar{c})$ data include moving boundary electrophoresis (Cann and Klapper 1961, Decker and Foster 1966), velocity sedimentation (Steinberg and Schachman 1966) and gel chromatography (Hummel and Dreyer 1962, Nichol and Winzor 1964). In general, the aim of these mass migration experiments is to provide a region of ligand alone, from which $\bar{c}$ may be determined. It is important to note, however, that generation of a region containing ligand alone does not necessarily ensure that the concentration within this region is $\bar{c}$. For example, in the simplest case of complex formation, viz. $A+S \rightleftharpoons AS$, it is possible (for example, in velocity sedimentation) that the relative order of magnitude of velocities of the individual entities would be $v_{AS} > v_A > v_S$. In this case, a trailing plateau of ligand could form, but the concentration determined from this plateau would be *less* than that of the unbound ligand in the original equilibrium mixture (see Chapter 4). Clearly, regardless of any precautions adopted to ensure an accurate estimation of this plateau concentration of ligand, the value would be inappropriate to the direct determination of meaningful binding results. It is a major aim of the next chapter to elucidate the requirements which must be fulfilled before binding data can be obtained from mass migration experiments on interacting systems.

The binding results, once obtained, may be plotted directly (r vs. $\bar{c}$; a binding curve), or in double-reciprocal, Scatchard, or Hill (1910) form. It is now clear that even in the simplest case, where the latter plots are linear, interpretation of the slopes and intercepts in terms of parameters relevant to the system must proceed with caution. Thus, while binding of ligand to equivalent and independent sites on the acceptor is indicated, the following possibilities arise: (i) a single acceptor entity binding a ligand which *either* does not self-interact (eqn (1.13)) *or* isomerizes (eqn (1.14)); (ii) a series of isomeric or polymeric forms of acceptor, with no sites lost in the self-interaction of acceptor and all intrinsic binding constants identical ($q=np$, $K_A=K_C$); and (iii) two isomeric forms of the acceptor both of which bind ligand at one site ($p=q=1$), regardless of whether $K_A=K_C$ or $K_A \neq K_C$ (eqn (1.15)). Similarly, it is unwise to interpret deviations from linearity in (say) the double-reciprocal plot in terms of one arbitrarily selected model. A curve convex to the $1/\bar{c}$ axis may be attributed to negative cooperativity as described by Koshland, but it is also possible that it is due wholly or partly to any or a combination of the following effects: (i) binding at non-equivalent sites; (ii) binding which is non-competitive with acceptor self-interaction (Fig. 1.4); and (iii) polymerization of the ligand, with preferential binding of the monomeric form. A double-reciprocal plot concave to the $1/\bar{c}$ axis (or a non-linear Hill plot) does not necessarily imply positive cooperativity (a special form of dependence of binding sites). An alternative model which leads to this behaviour (associated with sigmoidal binding curves) has been illustrated in Fig. 1.1 for the case of binding of ligand to sites equivalent and independent within each acceptor entity, the binding being preferential to one form of the acceptor, and competitive with the acceptor self-interaction. Fig. 1.3(c) illustrates another possibility, involving a self-interacting ligand. The ambiguity which arises in attempts to interpret binding results alone in terms of a microscopic model is not surprising, since the binding data are themselves macroscopic observations: information about the final equilibrium position cannot provide details of the pathway by which the equilibrium has been reached. It is true that dependence (or lack of it) of the forms of binding curves on $\bar{c}_{\text{P}}$ may indicate whether or not acceptor polymerization is involved (Fig. 1.1); but additional studies are required to interpret fully the results obtained. The obvious recommendation is to study the acceptor and ligand separately in order to characterize any self-interaction in terms of n, M_{A}, and X, and m, M_{S}, and Y (cf. eqn (1.1)): more complicated self-interaction behaviour may, of course, be indicated by such studies.

The determination of number-, weight-, and z-average molecular weights as a function of total solute (acceptor or ligand) concentration provides an excellent means of characterizing a system not only with respect to certain types of self-interaction, but also in terms of thermodynamic non-ideality. This approach, which has been reviewed extensively (for example, Nichol *et al.* 1964, Adams 1967), will not be emphasized in this book, which is concerned primarily with the use of commonly employed mass migration experiments for the study of interacting systems. We shall find that quantitative interpretation of mass migration results is possible in cases where polymerization (Chapter 3) and other types of self-interaction (Chapters 2 and 4) are encountered. Moreover, the methods are potentially valuable for the study of an isomerizing system, where the lack of concentration dependence of average molecular weights (as with a non-self-interacting solute) provides little information. The choice of environmental conditions for the study of the self-interaction of (say) the acceptor alone by any method should not be restricted to those selected for the determination of binding results. It is clear, for example, from Fig. 1.1 that a markedly sigmoidal binding curve may be associated with a large initial excess of non-binding polymer, so that in the absence of ligand or an effector acting as an activator (eqn (1.29)) the amount of active monomer may be undetectably small. However, even in the absence of ligands, variation of parameters which may effect X (such as temperature, ionic strength, or pH) could lead to detectable amounts of monomer coexisting in equilibrium with polymer. Thus, a study of β-A lactoglobulin at pH 4·65, ionic strength 0·1 and 20° would reveal essentially a single species with molecular weight 36 000, while at 4° in the same environment a rapidly equilibrating mixture of this and its tetrameric form would be observed (Townend, Winterbottom, and Timasheff 1960). Similarly, in the case of D-glyceraldehyde-3-phosphate dehydrogenase in buffer of low ionic strength (pH 7) only the tetrameric form was indicated at 20°, but lowering of the temperature promoted dissociation (Hoagland and Teller 1969). In both cases determination of X at different temperatures would lead to evaluation of ΔH^0 (from a plot of $\ln X$ vs. $1/T$) and hence to the calculation of X appropriate to the temperature used in binding studies. As we have noted in sections 4 and 5 of this chapter, the separate and joint effects of ligands, whether they be substrates, cofactors, or effectors, should be examined, particularly in the environment selected for binding studies.

The prediction in this chapter of sigmoidal binding curves is probably the most interesting observation in a biological context relating to metabolic control. Although emphasis has been given to isomerization and/or polymerization of acceptor and/or ligand as a basis of these effects, a further generalization may be made. *Basically, a sigmoidal binding curve may arise if ligand binds preferentially to at least one form of an acceptor which coexists in equilibrium with less active forms.** Certainly, these forms may be different conformational states, or polymers; but they might also result from reactions of the types $n\mathrm{A} + m\mathrm{B} \rightleftharpoons \mathrm{A}_n\,\mathrm{B}_m$ or $\mathrm{A} + \mathrm{B} \rightleftharpoons \mathrm{C} + \mathrm{D}$. It is possible, for example, that certain isoenzymes may arise via hybridization reactions of the latter type. Mass migration methods, notably zonal electrophoresis in a supporting medium, are generally used to characterize isoenzymes, and yet scant regard is paid to the complications which would arise in the final pattern in the event that a rapid equilibrium existed between the hybrid forms (see Chapter 4). It has also been noted (Masters, Sheedy, Winzor, and Nichol 1969) that an enzyme existing free in solution and bound to a matrix (i.e. $\mathrm{A} + \mathrm{B} \rightleftharpoons \mathrm{AB}$, with B (say) the cell membrane) may exhibit different activities towards a substrate, and that this may provide the basis of certain forms of metabolic control: a mechanism of this type has been suggested for aldolase in brain tissue (Clarke, Masters, and Winzor 1970). In this connection, any factor such as a drug,

* Preferential binding of ligand to one form of acceptor may be expressed by the conditions $q \neq np$ and/or $K_\mathrm{A} \neq K_\mathrm{C}$. Of this set, the possible gain or loss of binding sites on polymerization ($q \neq np$) has been mentioned: but, as noted earlier (Nichol *et al.* 1967a), it is also feasible that on polymer formation not directly involving the active sites the conformation of the constitutive monomer units may be perturbed in a way affecting the active sites ($q = np$ and $K_\mathrm{A} \neq K_\mathrm{C}$). An example is provided by the binding of certain competitive inhibitors to α-chymotrypsin (Nichol, Jackson and Winzor, 1972a). Conceptually, this mechanism differs from the proposals of ligand-induced conformational changes or ligand-induced association-dissociation reactions (Koshland, 1970), a difference which may be important in relation to enthalpy-entropy compensations in such systems (Lumry and Rajender, 1970).

hormone, metabolite, or a pH change in the vicinity of the membrane, which affected the equilibrium constant governing the $A + B \rightleftharpoons AB$ reaction would act as a physiological 'trigger' in governing the form of the sigmoidal response. The situation parallels that described in eqn (1.29) where an effector (possibly a metabolite acting as a feedback inhibitor) effectively modifies X: we note, however, that isomerization and polymerization constants may also be functions of other variables such as pH, ionic strength, pressure or temperature. In addition, it may be recalled from Fig. 1.2 that factors varying the constants describing ligand self-interaction may also operate to govern the form of a sigmoidal response. Other models implicating membranes in control phenomena may also be formulated. Changeux, Thiéry, Tung, and Kittel (1967) visualized two protomer globular lipoprotein entities of different conformations existing in equilibrium, with each capable of binding a single molecule of ligand with a different binding constant. With the appropriate change of symbols and placing $n=1$, $p=q=1$, $m_T=0$, their binding expression follows directly from eqn (1.11), except that the ratio m_C/m_A, which in one-phase systems is the isomerization constant, was replaced by a parameter which itself was a function of the state function, $\bar{m}_A/(\bar{m}_A+\bar{m}_C)$. Numerous examples showed that the binding curves deviated from the form of a rectangular hyperbola, the result expected on the basis of eqn (1.15) with X considered constant. Instead, a 'graded' (sigmoidal) response or an 'all-or-none' response was predicted.

In summary, there appears to be little doubt that binding studies will continue to occupy many workers, whose interests encompass control at the enzyme level, protein synthesis, the action of pharmacological agents and hormones, and many other fields. It is also highly probable that many systems will be encountered in which acceptor or ligand (or both) self-interacts, and that in some instances these interactions may provide the basis of important biological effects. Regardless of the validity of the latter prediction, it is certain that the interactions will lead to complications in the experimental records of migration experiments. The recognition of this complicated behaviour inherent in migration of interacting systems, and the use of the migration patterns to characterize the interactions are the concerns of the remaining chapters.

2

THE USE OF CONSTITUENT QUANTITIES AND CONTINUITY EQUATIONS TO DESCRIBE MIGRATION PATTERNS

2.1. Migration experiments involving a plateau region of original composition

THREE types of mass migration experiment have been commonly employed for the study of interacting systems, namely, electrophoresis, sedimentation, and gel chromatography. Despite the vast differences between the equipment and technical operations inherent in these techniques, they have in common the ability to impart characteristic rates of migration to individual solutes: the origins of these velocities do, of course, differ. In electrophoresis separations of solutes are effected by virtue of different responses of solutes to an electric field, the equivalent force in sedimentation being a gravitational (centrifugal) field: partitioning of solute between the mobile and stationary phases governs migration in chromatography. Two different procedures have been used in each case for measuring these velocities, viz. zonal and frontal analysis.

In zonal analysis a small volume (or zone) of solute is introduced at the top of the column, be it a gel bed or a sucrose gradient. The simplest theoretical treatment of zonal migration requires the concentration of solute to be uniform throughout the initial zone, and zero elsewhere. On applying the relevant field (electric, centrifugal, or elution with solvent) the zone rapidly loses its discontinuous character, and assumes a Gaussian form in the case of a single, non-interacting solute. In this type of experiment the final pattern, usually a plot of concentration (or related parameter) vs. volume (or distance), exhibits a series of peaks or zones, which may be used to identify the velocities and amounts of solutes comprising a non-interacting mixture: Fig. 2.1(a) illustrates this form of pattern obtained with a mixture of non-interacting solutes A and B. If, however, a solution containing (say) polymeric species in rapid equilibrium were subjected to zonal analysis, a fundamental difficulty would arise because of the inevitable spreading and consequent dilution of the zone during migration (Schumaker and Rosenbloom 1965, Winzor 1966). The essentially instantaneous response of the monomer-polymer equilibria to a changed (diluted) environment would result in a progressive enrichment of the zone in terms of monomer, at the expense of polymer species, and a corresponding, continual change in the rate of migration. The consequent insensitivity and difficulty in interpreting zonal data on interacting systems

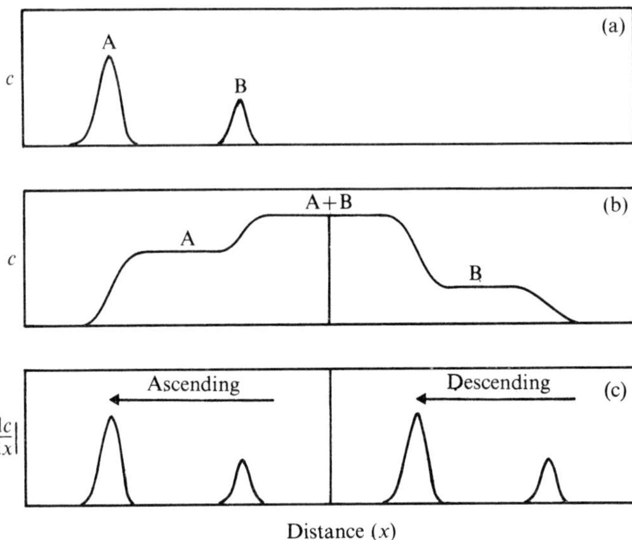

FIG. 2.1. Migration patterns for a mixture of two non-interacting solutes A and B: (a) zonal analysis; (b) frontal analysis (integral form); (c) frontal analysis (plotted in differential form).

are evident from the results of calculations which simulate the gel chromatographic behaviour of ovalbumin on Sephadex G-100 (Winzor 1966).

Investigations of chemically reacting systems are greatly facilitated by performing mass migration experiments in such a way that a plateau region of original composition is maintained throughout the run. For simplicity we again examine the behaviour of a mixture of two non-interacting solutes to demonstrate the features of frontal analysis (Fig. 2.1(b)). In this technique, plateau regions of solvent ($c=0$), single solutes (A and B), and the solution of original composition (A + B) separate boundary regions (concentration gradients of either A or B). Differentiation of this pattern leads to Fig. 2.1(c) which clearly shows that frontal analysis, in contrast with the zonal method, records conjugate ascending (advancing) and descending (trailing) sides. In this instance, where the absolute magnitude of the concentration gradient is plotted as the ordinate, the conjugate sides are mirror images, or *enantiographic* with respect to numbers, positions and areas of peaks.

Conjugate ascending and descending sides are observed in moving boundary electrophoresis conducted in a Tiselius U-tube cell assembly, where a plateau region of original composition is preserved throughout the experiment (Longsworth 1959*b*). In the usual design of velocity sedimentation, where a cell is filled initially with solution of uniform composition, only one boundary system is generated, viz. that in which solute migrates away from a solvent plateau. Although pile-up of material occurs in the vicinity of the cell base, a plateau region of original composition may be considered to be maintained by adopting the 'rectangular approximation', which assumes that the cell is uniform in cross-section (instead of sector-shaped) and that the applied field is homogeneous. These conditions of uniformity are realized much more closely in moving boundary electrophoresis and frontal chromatography. In the latter procedure, sufficient solution is added to the column to ensure that the elution profile contains a plateau region in which the concentration of solute(s) equals that applied. The resulting profile, in which the concentration of solute rises from zero to the original value, reflects the migration (or advance) of the plateau region into a region occupied previously by solvent. Subsequent elution of the column with solvent generates a conjugate trailing elution profile in which the concentration decreases from the applied value to zero.

In order to continue the comparison between the three different types of mass migration experiment it is necessary to assign a velocity per unit applied field to each solute species. The relevant parameter in moving boundary electrophoresis is clearly the electrophoretic mobility, U ($cm^2s^{-1}V^{-1}$), while the sedimentation coefficient, s (s) reflects velocity of migration in the ultracentrifuge. A detailed consideration of the correct analogue of velocity in gel chromatography will be delayed until section 4 of this chapter; but suffice it for the present to say that in the treatment of elution profiles the correct parameter is the partition coefficient (Ackers and Thompson 1965) or the elution volume (Gilbert 1966*a*; Nichol, Ogston, and Winzor 1967*b*).

Thus it is possible to write continuity equations which describe conservation of mass in any of the three migration processes involving a plateau region of original composition, and to deduce the forms of the boundaries which might arise in the study of an interacting system. We introduce the necessary concepts by considering first a single non-associating solute.

2.2. Continuity equation for a single non-associating solute

Consider a tube of uniform cross-sectional area S, along which the single solute is migrating axially: in so doing it crosses sections 1 and 2, which are perpendicular to the axis and situated at positions x and $x+\mathrm{d}x$ from some arbitrary origin (Fig. 2.2). The amount (weight) of solute crossing a section due to directed migration alone is in general given by the product of area (cm^2), velocity ($cm\ s^{-1}$), concentration ($g\ cm^{-3}$), and time interval (s) under consideration. In practice there will also of necessity be a diffusional flow of material whenever a gradient of chemical potential is encountered. A continuity equation may be written which includes a term to account for diffusional flow, but its solution is difficult even for this simplified case. Accordingly, in order to illustrate the basic forms of boundary distributions to a first approximation, we shall follow the procedure of Gilbert (1959), Gilbert and Jenkins (1959), and Nichol and Ogston (1965*a*, 1967) by ignoring initially all diffusional flows. Diffusional spreading is proportional to $\sqrt{t}$ whereas migration varies linearly with time, and consequently, solutions of the continuity equation without diffusional terms describe the shape of migration patterns that would be obtained at infinite time. A disadvantage of these asymptotic profiles is that it is frequently difficult at first encounter to ascertain whether experimental patterns are in agreement or disagreement with prediction. However, their use has furthered significantly our understanding of the mass

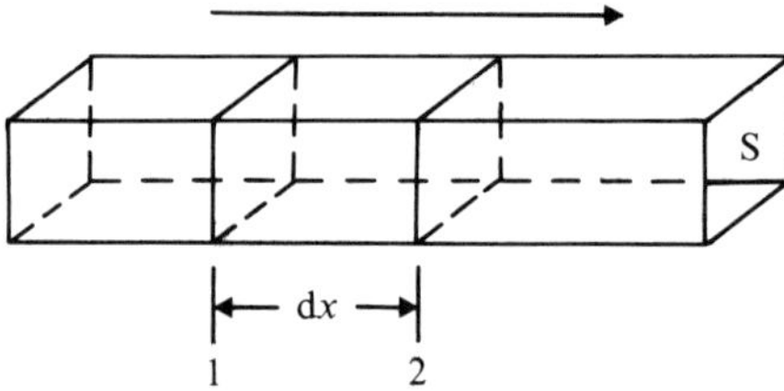

FIG. 2.2. Schematic representation of migration in a cell of uniform cross-section.

migration behaviour of interacting solutes, well in advance of the development of procedures for incorporating effects of diffusion: these will be discussed in Chapter 5. Furthermore, although the asymptotic profiles may not be observed directly in a single experiment, they may be constructed from results of a series of differential migration experiments (Gilbert 1966*b*; Gilbert and Gilbert 1968; Chiancone, Gilbert, Gilbert, and Kellett 1968).

In a time interval from t to $t+\mathrm{d}t$ the amounts of solute crossing sections 1 and 2 in Fig. 2.2 due to directed migration are as follows:

$$\text{section 1:}\quad Sv(x, t)c(x, t)\,\mathrm{d}t,$$

$$\text{section 2:}\quad Sv(x+\mathrm{d}x, t)c(x+\mathrm{d}x, t)\,\mathrm{d}t,$$

where the concentration and velocity of the solute are functions of x and t, but are uniform in any section of tube normal to its axis. The net quantity of solute entering the volume element $\mathrm{d}V$ $(=S\,\mathrm{d}x)$ in the selected time interval is given by the difference between the above two terms. A statement of conservation of mass is obtained by equating this difference to the variation in amount of solute within the volume element $(S\,\mathrm{d}x\cdot\mathrm{d}c)$ during the same time interval. Cancellation of S (stressing that the cell may be of any shape with uniform cross-section), and extrapolation to the limit of infinitessimally small intervals of time and distance yields eqn (2.1):

$$\frac{\partial c}{\partial t} = -\frac{\partial (vc)}{\partial x}. \tag{2.1}$$

If v is a function only of c this equation becomes

$$\frac{\partial c}{\partial t} + \frac{\mathrm{d}(vc)}{\mathrm{d}c}\frac{\partial c}{\partial x} = 0. \tag{2.2}$$

Eqn (2.2) describes in three-dimensional space a surface with cartesian coordinates x, t, c. The intersections of this surface with the planes described by

$$c(x, t) = \text{constant} \tag{2.3a}$$

or

$$\frac{\partial c}{\partial x}\mathrm{d}x + \frac{\partial c}{\partial t}\mathrm{d}t = 0 \tag{2.3b}$$

are given by the simultaneous solution of eqns (2.2) and (2.3b), viz.

$$\frac{\mathrm{d}x}{\mathrm{d}t} = \frac{\mathrm{d}(vc)}{\mathrm{d}c}. \tag{2.4}$$

Eqns (2.3a) and (2.4) define a family of straight lines parallel to the plane $c(x, t) = 0$. The initial boundary condition is now introduced that at zero time a perfectly sharp boundary separates solvent ($c = 0$) from original solution ($c = c_0$) plateaux, such that

$$c(x, 0) = 0 \quad \text{for all } x \leq 0,$$

$$c(x, 0) = c_0 \quad \text{for all } x > 0.$$

This now permits integration of eqn (2.2) for particular values of t to give expressions describing the concentration profile in the corresponding (x, c) plane. In this case it is convenient to follow the procedure adopted by Nichol and Ogston (1965*a*), who represented c as a function of the variable u ($u = x/t$), defined for all finite, non-zero values of t as the velocity of a lamina of constant composition. With this approach the continuity equation corresponding to eqn (2.4) may be written as

$$u\,\mathrm{d}c - \mathrm{d}(vc) = 0, \tag{2.5}$$

which describes the concentration profile in the (u, c) plane.

The simplest application of eqn (2.5) is consideration of a solute whose velocity is concentration-independent, i.e.

$$\mathrm{d}v/\mathrm{d}c=0 \text{ and } \mathrm{d}(vc)/\mathrm{d}c=v=\text{constant}.$$

Eqn (2.5) simplifies to $(u-v)\,\mathrm{d}c=0$, which has the solutions that *either* $\mathrm{d}c=0$ *or* $u=v$. The first solution describes the solvent ($c=0$) and solution ($c=c_0$) plateaux, while the second solution, viz. $u=v$, describes the single sharp boundary separating the plateaux and situated at a distance $vt=ut$ from the origin: the same solution applies to the ascending and descending sides, which would accordingly be enantiographic. In practice such boundaries, termed sharp by Nichol and Ogston (1965*a*, 1967) would be spread by diffusion. Involvement of the solute in either frictional or chemical interactions introduces concentration-dependence of migration rate ($\mathrm{d}v/\mathrm{d}c\neq0$), whereupon solution of eqn (2.5) requires specification of the relationship between velocity and concentration. Examples of chemical interactions of biological significance have already been cited in Chapter 1, while the s–c dependence observed with most biological macromolecules (for example, proteins, nucleic acids, and polysaccharides) in velocity sedimentation serve to illustrate the relevance of physical effects: both of these types of phenomena will be treated, the former first since the functional relationships between velocities and concentrations are governed by the law of mass action.

2.3. Constituent concentrations and velocities

In the case of a chemically interacting system subjected to mass migration, a continuity equation similar to eqn (2.5) may be written provided allowance is made for the fact that any chemically definable species A may exist in more than one state. For example, solute A may exist as monomeric and polymeric forms ($n\mathrm{A}\rightleftharpoons\mathrm{C}$) or as unreacted A and in combination with another reactant (for example, $\mathrm{A}+\mathrm{B}\rightleftharpoons\mathrm{AB}$). Thus we are required to consider the constituent concentration and constituent velocity of A when writing a continuity equation to describe the conservation of its mass. The constituent concentration of A in a particular region of the cell is its total concentration irrespective of the state in which it exists. Each state of A has a characteristic velocity, and the constituent velocity is a weighted-average velocity, taken over all states of A coexisting in a particular region of the cell at any given time (Tiselius 1930, Longsworth 1959*a*, Nichol and Ogston 1965*a*). Particular examples will now be presented to illustrate these definitions and to provide equations for later use. For the present we shall consider only those situations in which the rate of interchange between states is sufficiently rapid for the constituent velocity to be time-independent.

(a) *A polymerization or isomerization* ($n\mathrm{A} \rightleftharpoons \mathrm{C}$)

The constituent concentration ($\bar{c}$) and velocity ($\bar{v}$) of this single, self-interacting solute are given by

$$\bar{c} = c_{\mathrm{A}} + c_{\mathrm{C}}, \tag{2.6a}$$

$$\bar{v} = \frac{(v_{\mathrm{A}} c_{\mathrm{A}} + v_{\mathrm{C}} c_{\mathrm{C}})}{\bar{c}}, \tag{2.6b}$$

where c_{A} and c_{C} are expressed on a g/unit volume scale: $\bar{v}$ is thus the weight-average velocity. In what follows, the velocities of individual species (states) are considered to be concentration-independent, and thus $\bar{c}$ and $\bar{v}$ are interrelated solely by the law of mass action, expressed as

$$K = \frac{c_{\mathrm{C}}}{c_{\mathrm{A}}^{n}} \tag{2.7a}$$

or

$$n(c_{\mathrm{A}})^{n-1} K \, \mathrm{d}c_{\mathrm{A}} = \mathrm{d}c_{\mathrm{C}}. \tag{2.7b}$$

The derivation of eqn (2.4) given in the previous section remains valid when written in terms of these constituent quantities, since it merely states the requirement of conservation of mass. Accordingly, the appropriate equation of continuity is

$$u \, \mathrm{d}\bar{c} - \mathrm{d}(\bar{v}\bar{c}) = 0. \tag{2.8}$$

(b) *A heterogeneous association reaction of the type* $\mathrm{A} + \mathrm{B} \rightleftharpoons \mathrm{C}$

A set of any two constituent quantities suffices to describe the system: we choose

$$\bar{m}_{\mathrm{A}} = m_{\mathrm{A}} + m_{\mathrm{C}}, \tag{2.9a}$$

$$\bar{m}_{\mathrm{B}} = m_{\mathrm{B}} + m_{\mathrm{C}}, \tag{2.9b}$$

$$\bar{v}_{\mathrm{A}} = \frac{(v_{\mathrm{A}} m_{\mathrm{A}} + v_{\mathrm{C}} m_{\mathrm{C}})}{\bar{m}_{\mathrm{A}}}, \tag{2.9c}$$

$$\bar{v}_{\mathrm{B}} = \frac{(v_{\mathrm{B}} m_{\mathrm{B}} + v_{\mathrm{C}} m_{\mathrm{C}})}{\bar{m}_{\mathrm{B}}}, \tag{2.9d}$$

where m_{A} and m_{B}, etc., are expressed in moles (or base moles) per unit volume: $\bar{v}_{\mathrm{A}}$ and $\bar{v}_{\mathrm{B}}$ are clearly no longer weight-average velocities. The association equilibrium constant may be written as

$$K = \frac{m_{\mathrm{C}}}{m_{\mathrm{A}} m_{\mathrm{B}}} \tag{2.10}$$

and the continuity equations as

$$u_{\rm A}\ {\rm d}\bar{m}_{\rm A}-{\rm d}(\bar{v}_{\rm A}\bar{m}_{\rm A})=0, \tag{2.11a}$$

$$u_{\rm B}\ {\rm d}\bar{m}_{\rm B}-{\rm d}(\bar{v}_{\rm B}\bar{m}_{\rm B})=0. \tag{2.11b}$$

(c) *A hybridization reaction* (A + B ⇌ C + D)

This reaction may be written quite generally as $p_1p_2+q_1q_2 \rightleftharpoons p_1q_1+p_2q_2$, whereupon the following equations may be written in terms of the equilibrium concentrations $m_{\rm A}$, $m_{\rm B}$, $m_{\rm C}$, and $m_{\rm D}$ (expressed once again on a molar scale).

$$\bar{m}_{p1}=m_{\rm A}+m_{\rm C};\quad \bar{v}_{p1}\bar{m}_{p1}=v_{\rm A}m_{\rm A}+v_{\rm C}m_{\rm C}; \tag{2.12a}$$

$$\bar{m}_{p2}=m_{\rm A}+m_{\rm D};\quad \bar{v}_{p2}\bar{m}_{p2}=v_{\rm A}m_{\rm A}+v_{\rm D}m_{\rm D}; \tag{2.12b}$$

$$\bar{m}_{q1}=m_{\rm B}+m_{\rm C};\quad \bar{v}_{q1}\bar{m}_{q1}=v_{\rm B}m_{\rm B}+v_{\rm C}m_{\rm C}; \tag{2.12c}$$

$$\bar{m}_{q2}=m_{\rm B}+m_{\rm D};\quad \bar{v}_{q2}\bar{m}_{q2}=v_{\rm B}m_{\rm B}+v_{\rm D}m_{\rm D}; \tag{2.12d}$$

$$K=\frac{m_{\rm C}m_{\rm D}}{m_{\rm A}m_{\rm B}}; \tag{2.13}$$

$$u_Y\ {\rm d}\bar{m}_Y-{\rm d}(\bar{v}_Y\bar{m}_Y)=0; \tag{2.14}$$

where the subscript Y in the continuity expression (eqn (2.14)) represents the general constituent (p_1, p_2, q_1, or q_2). It is clear that only three constituents need to be considered in order to describe the system, since any three of the expressions listed in eqn (2.12) define the fourth.

Having illustrated the notion of constituent quantities with these examples we are now in a position to utilize the continuity equations containing them to deduce the forms of migrating boundaries, employing the approach adopted by Nichol and Ogston (1965*a*, 1967): this assumes that $u_{\rm A}$ and $u_{\rm B}$ in eqns (2.11a) and (2.11b) (for example) exist and are identical. But before proceeding with the deduction of boundary shapes let us examine more closely these constituent concepts of concentration and of velocity as they apply in chromatography.

2.4. Analogues of concentration and velocity in chromatography

Although frontal gel filtration, moving boundary electrophoresis, and velocity sedimentation may be grouped together under the heading of mass migration techniques involving plateaux, the presence of the stationary phase in chromatography does distinguish it from the two freely-migrating systems:

indeed all solute and solvent molecules possess the same velocity, namely the flow rate of the column. Retardation of a solute A reflects an interaction between A and the station phase G, which may be represented as

$$\mathrm{A} + G \rightleftharpoons \mathrm{A}G, \tag{2.15}$$

the complex having zero velocity. Provided that this interphasal equilibration is very rapid compared with solute flow, the net migration of a single solute in chromatography may be regarded as a constituent velocity, determined by the position of the equilibrium in eqn (2.15). Similarly, in order to describe the movement of such a solute within the column it is necessary to take into account the concentrations of A in mobile and stationary phases. Thus description of the chromatographic behaviour of chemically reacting systems requires modification of the theoretical treatment developed for free migration by introducing the concept of inclusive constituent quantities ($\bar{\bar{c}}$ or $\bar{\bar{m}}_\mathrm{A}$; and $\bar{\bar{v}}$ or $\bar{\bar{v}}_\mathrm{A}$, etc.). For systems in rapid chemical equilibrium the constituent quantities $\bar{c}$ (or $\bar{m}_\mathrm{A}$) and $\bar{v}$ (or $\bar{v}_\mathrm{A}$), as defined earlier, refer to the mobile phase of the column, whereas $\bar{\bar{c}}$ (or $\bar{\bar{m}}_\mathrm{A}$) and $\bar{\bar{v}}$ (or $\bar{\bar{v}}_\mathrm{A}$) also take into account the interaction between mobile and staionary phases (eqn (2.15)). The appropriate equations of continuity describing the process of chromatographic transport as a whole are then

$$u\,\mathrm{d}\bar{\bar{c}} - \mathrm{d}(\bar{\bar{v}}\bar{\bar{c}}) = 0 \quad \text{(single associating solute)} \tag{2.16a}$$

or

$$u_\mathrm{A}\,\mathrm{d}\bar{\bar{m}}_\mathrm{A} - \mathrm{d}(\bar{\bar{v}}_\mathrm{A}\bar{\bar{m}}_\mathrm{A}) = 0 \quad \text{(heterogeneous} \tag{2.16b}$$

$$u_\mathrm{B}\,\mathrm{d}\bar{\bar{m}}_\mathrm{B} - \mathrm{d}(\bar{\bar{v}}_\mathrm{B}\bar{\bar{m}}_\mathrm{B}) = 0 \quad \text{association).} \tag{2.16c}$$

There now remains the problem of obtaining expressions for these inclusive constituent parameters, which we shall illustrate by considering a reversibly dimerizing system ($2\mathrm{A} \rightleftharpoons \mathrm{C}$).

If the position of the partition equilibrium of monomer (A) is defined by μ_A, the fraction of A that is in the mobile phase, then the constituent concentration of A within the column, $\bar{c}_\mathrm{A}^J$, is given by the relationship

$$\bar{c}_\mathrm{A}^J = \frac{\phi c_\mathrm{A}}{\mu_\mathrm{A}}, \tag{2.17}$$

where ϕ denotes the fraction of cross-sectional area of column that is occupied by mobile phase: the corresponding expression for $\bar{\bar{c}}$ is then given by eqn (2.18),

$$\bar{\bar{c}} = \bar{c}_\mathrm{A}^J + \bar{c}_\mathrm{C}^J, \tag{2.18}$$

which on combination with eqns (2.7a) and (2.17) yields

$$\bar{\bar{c}} = \phi\left(\frac{c_{\mathrm{A}}}{u_{\mathrm{A}}} + \frac{Kc_{\mathrm{A}}^2}{u_{\mathrm{C}}}\right). \qquad (2.19)$$

$\bar{v}_{\mathrm{A}}^{I}$ and $\bar{\bar{v}}$ are constituent velocities corresponding with the definitions of $\bar{c}_{\mathrm{A}}^{I}$ and $\bar{\bar{c}}$, noting, however, that v_{A} and $\bar{v}$ are identical with v, the linear velocity of the mobile phase. Since the velocities of AG and CG are zero, it follows that

$$\bar{\bar{v}} = \frac{\phi\bar{v}\bar{c}}{\bar{\bar{c}}} = \frac{\phi v\bar{c}}{\bar{\bar{c}}}. \qquad (2.20)$$

In a column of unit cross-sectional area and length l, through which the solvent flow rate is $\dot{V}$ (equals ϕv), the time taken for a lamina moving with velocity u is l/u, the corresponding volume of eluate, V_u, being given by

$$V_u = \frac{\dot{V}l}{u}. \qquad (2.21)$$

Thus, as expected, there is a reciprocal relationship between velocity within the column and elution volume; accordingly $1/V$ is the analogue of v when inclusive constituent concentrations are recorded. However, such concentrations are not usually obtained, unless direct column scanning (Brumbaugh and Ackers 1968, 1971) is used to follow chromatographic migration: the more conventional experimental record is an elution profile.

Conservation of total solute across the entry plane of the column requires that

$$\dot{V}\bar{c}' = \bar{\bar{v}}\bar{\bar{c}}, \qquad (2.22)$$

where $\bar{c}'$ denotes the constituent concentration of A in the inflowing solution. Substitution of eqn (2.20) for $\bar{\bar{v}}$ yields the identity

$$\bar{c}' \equiv \bar{c}. \qquad (2.23)$$

This relationship also applies at the exit plane of the column, and it therefore follows that all constituent concentrations of solutes in the eluate at any moment may be identified with those in the mobile phase solution which reaches the exit plane at that time: the profile of composition in the elution profile therefore exactly reproduces that in the mobile phase of a moving boundary system. However, the equation of continuity (eqn (2.16)) refers to the composition of the boundary as a whole, and must accordingly be modified if it is to describe only the mobile phase.

Since we wish to discuss the shape of the elution profile, conversion from the parameter u to its elution counterpart, V_u, is clearly desirable. After substituting eqns (2.20) and (2.21) in the continuity expression (eqn (2.16a)) we find that

$$V_u = \frac{\dot{V}\,\mathrm{d}\bar{\bar{c}}}{\phi v\,\mathrm{d}\bar{c}} = \frac{V_o\,\mathrm{d}\bar{\bar{c}}}{\phi\,\mathrm{d}\bar{c}}, \tag{2.24}$$

where V_o, the void volume of the column, is the volume occupied by mobile phase. Application of this general expression to a monomer–dimer system by inclusion of the relevant terms for $\bar{c}$ and $\bar{\bar{c}}$ (eqns (2.7) and (2.19)) leads to the relationship

$$\left(V_u - \frac{V_o}{\mu_A}\right)\mathrm{d}c_A + 2Kc_A\left(V_u - \frac{V_o}{\mu_C}\right)\mathrm{d}c_A = 0. \tag{2.25}$$

Consideration of the chromatographic behaviour of a single, non-interacting solute X shows that $V_X = V_o/\mu_X$, whereupon eqn (2.25) may be written as

$$(V_u - V_A)\,\mathrm{d}c_A + 2Kc_A(V_u - V_C)\,\mathrm{d}c_A = 0. \tag{2.26}$$

Combination of eqns (2.6)–(2.8), referring to migration in the absence of a stationary phase, yields an identical expression except that velocities replace elution volumes. Thus adaptation of theoretical relationships in terms of velocities to elution profiles merely requires this direct substitution of elution volumes (Gilbert 1966*a*, Nichol *et al.* 1967*b*): suitably defined partition coefficients also fulfil this purpose (Ackers and Thompson 1965).

2.5. Deduction of the form of migration patterns relevant to the evaluation of binding data

Even in cases where constituent velocities are functions only of composition and not of time (for example, where all relevant chemical equilibria are rapidly attained), many solutions of the continuity equation are possible for both the ascending and descending sides, depending on (a) the nature of the interactions involved and (b) the relative order of magnitude of the velocities of the individual species present. Many examples of different solutions will be given in Chapters 3 and 4; but let us commence by illustrating the Nichol and Ogston (1965*a*) approach with a system of practical significance in evaluating binding data discussed in Chapter 1.

(a) *The model and basic equations*

Consider a solution containing a self-interacting acceptor ($n\mathrm{A} \rightleftharpoons \mathrm{C}$) and a self-interacting ligand ($m\mathrm{S} \rightleftharpoons \mathrm{T}$), as described in Chapter 1. The system is allowed to reach equilibrium, and is subjected to frontal gel chromatography on (say) a Sephadex column, pre-equilibrated with solvent. It is assumed that there is no irreversible adsorption of any solute to the stationary phase. In addition, the velocity of each solute species on the column is assumed to be constant and independent of the solution composition in its local environment: these velocities are denoted by v_A, v_C, v_S, v_T, $v_{AS_iT_j}$, and $v_{CS_kT_l}$, where $i=0, 1, 2, \ldots p$, $j=0, 1, 2, \ldots p$ $(0 \leqq (i+j) \leqq p)$, and $k=0, 1, 2, \ldots q$, $l=0, 1, 2, \ldots q$ $(0 \leqq (k+l) \leqq q)$. Suppose now we envisage that a stationary phase has been chosen which excludes A, C, and all complexes AS_iT_j and CS_kT_l, so that each of these species migrates down the column with the velocity of the mobile phase: the unbound ligands S and T interact with the gel (eqn (2.15)) and thus move with slower velocities. (Strictly speaking we should refer to constituent velocities in chromatography, but we shall adhere to the terminology for freely migrating systems in this and all future derivations of mass migration patterns.) The situation may be summarized as

$$v_S < v_T < v_A = v_C = v_{AS_iT_j} = v_{CS_kT_l}. \tag{2.27}$$

The constituent concentrations of acceptor ($\bar{c}_P$ g/litre) and ligand ($\bar{c}_L$ g/litre) are given by eqns (1.9) and (1.10), respectively, provided that binding is to equivalent and independent binding sites within each acceptor entity. The corresponding constituent velocities, $\bar{v}_P$ and $\bar{v}_L$, may be defined using condition (2.27) by

$$\bar{v}_P \bar{c}_P = M_A\{v_A m_A(1 + K_A m_S + L_A m_T)^p + \\ + v_A n m_C(1 + K_C m_S + L_C m_T)^q\}, \tag{2.28a}$$

$$\bar{v}_L \bar{c}_L = M_S\{v_S m_S + m v_T m_T + v_A p m_A \\ (1 + K_A m_S + L_A m_T)^{p-1}(K_A m_S + m L_A m_T) \\ + v_A q m_C(1 + K_C m_S + L_C m_T)^{q-1}(K_C m_S + m L_C m_T)\}, \tag{2.28b}$$

The appropriate continuity equations are

$$u \, \mathrm{d}\bar{c}_P - \mathrm{d}(\bar{v}_P \bar{c}_P) = 0, \tag{2.29a}$$

$$u \, \mathrm{d}\bar{c}_L - \mathrm{d}(\bar{v}_L \bar{c}_L) = 0. \tag{2.29b}$$

Combination of eqns (1.9) and (2.28a) shows that with condition (2.27) $\bar{v}_P = v_A =$ constant, and thus eqn (2.29a), written as $(u - \bar{v}_P)\, \mathrm{d}\bar{c}_P - \bar{c}_P\, \mathrm{d}\bar{v}_P = 0$ simplifies to

$$(u - v_A)\, \mathrm{d}\bar{c}_P = 0. \tag{2.30}$$

The corresponding expression for ligand conservation is obtained by combining eqns (1.10) and (2.28b) in differential form, and (2.29b) to give

$$(u-v_S)\,\mathrm{d}m_S+m(u-v_T)\,\mathrm{d}m_T$$
$$+(u-v_A)D\{pm_A(K_A m_S+mL_A m_T)(1+K_A m_S+L_A m_T)^{p-1}$$
$$+qm_C(K_C m_S+mL_C m_T)(1+K_C m_S+L_C m_T)^{q-1}\}=0. \quad (2.31)$$

The operation of differentiation indicated in this equation by $D\{\quad\}$ may be performed; but the equation is left in unexpanded form for the sake of simplicity.

The required set of basic equations is completed by examination of the equilibrium conditions,

$$X=\frac{m_C}{m_A^n};\quad \frac{nm_C}{m_A}\,\mathrm{d}m_A=\mathrm{d}m_C; \quad (2.32a)$$

$$Y=\frac{m_T}{m_S^m};\quad \frac{mm_T}{m_S}\,\mathrm{d}m_S=\mathrm{d}m_T. \quad (2.32b)$$

Equation (2.32a) shows that *either* $\mathrm{d}m_A=\mathrm{d}m_C=0$, *or* they are of the same sign: eqn (2.32b) shows that the same possibilities apply to $\mathrm{d}m_S$ and $\mathrm{d}m_T$. Moreover, while either A and C or S and T may coexist in a region, the existence of all four would require the additional presence of all complexes, AS_iT_j and CS_kT_l.

In what follows, reference is made to Fig. 2.3, which presents the concentration–velocity profile for both ascending and descending sides for the system under discussion. The values of the equilibrium concentrations of A, C, S, T, all AS_iT_j, and all CS_kT_l in the original solution plateau have been chosen quite arbitrarily: in practice, different values most certainly pertain to the individual complexes, but the important features of the elution pattern are not obscured by summing the concentrations of all complexes AS_iT_j and separately of CS_kT_l, and representing them as such in Fig. 2.3.

(b) *The solution for the descending side*

The convention is adopted that directions (and hence $u=x/t$) are positive in the direction solvent→solution.

(i) *The plateau of original composition.* When $u>v_A$, then $(u-v_A)\neq 0$ and eqn (2.30) is satisfied only by $\mathrm{d}\bar{c}_P=0$. This implies that there is no change in concentrations of A, C, and complexes, and hence no change in the concentrations of S and T. Eqn (2.31) is also satisfied by placing the differentials of all species concentrations equal to zero. In frontal analysis these relations must describe the plateau of original composition, wherein the equilibrium concentrations, m_S, m_T, m_A, m_C, all $m_{AS_iT_j}$, and all $m_{CS_kT_l}$ are non-zero.

(ii) *The position* $x = v_A t$. When $u = v_A$, then $(u - v_A) = 0$ and eqn (2.30) allows *either* $d\bar{c}_P = 0$ *or* $d\bar{c}_P \neq 0$. In fact the latter condition must pertain, since it describes a change in concentration of all forms of acceptor and their

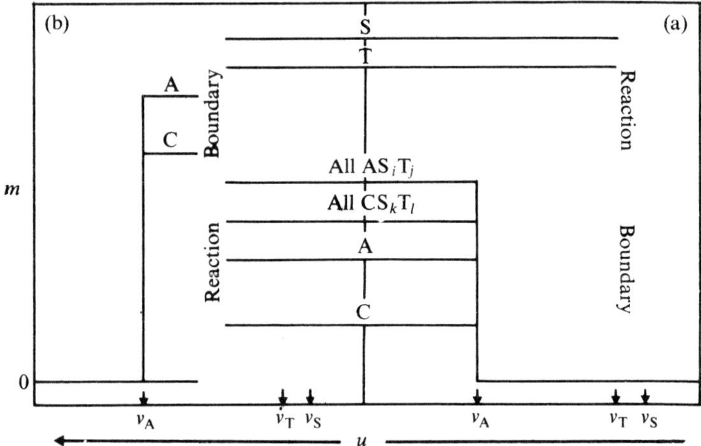

FIG. 2.3. A representative plot of concentration against the normalized distance parameter, u, for the system

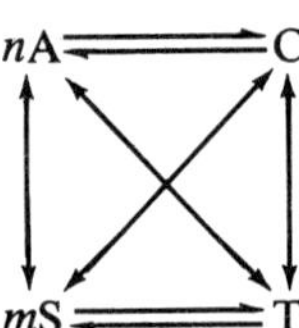

where $\longleftrightarrow$ denotes multiple binding of ligand to form a series of complexes AS_iT_j and CS_kT_l (see text). Values assigned to individual species concentrations and velocities (marked on the abscissa) are completely arbitrary except that the condition $v_S < v_A = v_C =$ all $v_{AS_iT_j} =$ all $v_{CS_kT_l}$ holds. Distances are positive in the direction solvent$\rightarrow$solution (descending side) and solution$\rightarrow$solvent (ascending side).

complexes which must occur on continued elution, and can only occur when $u = v_A$:* thus, from eqn (2.30) $d\bar{c}_P = 0$ when $u \neq v_A$. It also follows that the

* This conclusion may be confirmed by noting that the median bisector of the acceptor constituent gradient in this boundary region is given by (Nichol and Ogston 1965*a*, *b*)

$$\bar{u} = \frac{\int u \, d\bar{c}_P}{\int d\bar{c}_P} = \frac{\int d(\bar{v}_P \bar{c}_P)}{\int d\bar{c}_P} = v_A,$$

the integrations being taken from the original solution plateau to one containing only S and T.

changes in m_A, m_C, all $m_{AS_iT_j}$, and all $m_{CS_kT_l}$ at $x=v_A t$ are from the original plateau values to zero, as shown in Fig. 2.3, descending side. At the same position, when $(u-v_A)=0$, eqn (2.31) becomes

$$(u-v_S)\,dm_S+m(u-v_T)\,dm_T=0. \tag{2.33}$$

Since at this position $(u-v_S)$ and $(u-v_T)>0$, the only solution of eqn (2.33) is $dm_S=dm_T=0$, because dm_S and dm_T cannot be of opposite sign (eqn (2.32b)). The important conclusion is reached that there is no change in the concentrations of S and T from their original equilibrium values at the position $x=v_A t$.

(iii) *The region* $v_T<u<v_A$. In this region $(u-v_A)$, $(u-v_S)$, and $(u-v_T)\neq 0$, with the result that the only solution of both continuity expressions (eqns (2.30) and (2.31)) is that the differentials of the concentrations of all entities are zero. Hence, a plateau region extends containing only S and T *at their equilibrium concentrations in the original mixture.* Indeed, if m_A and m_C are equated to zero in eqn (1.10) it follows that in this region

$$\bar{c}_L=M_S(m_S+mm_T). \tag{2.34}$$

Thus, if the total concentration of ligand in this region were measured, the value would be the total equilibrium concentration of all forms of *unbound* ligand in the original equilibrium mixture, i.e. it would be $\bar{c}$ as defined and employed in the discussion of binding results presented in Chapter 1.

(iv) *The region* $v_S<u<v_T$. Since $m_A=m_C=0$ in this region, it follows from eqn (2.31) that eqn (2.33) applies. In the region under discussion $(u-v_S)$ and $(u-v_T)\neq 0$ and are of opposite sign. The solution of eqn (2.33), therefore, requires that dm_S and $dm_T\neq 0$, and that both differentials have the same sign, which is in accord with eqn (2.32b). These changes in concentration of S and T, which can only occur in this region, must be from their original values to zero. Thus, when $u<v_S$, the continuity equations are only satisfied by the condition that the differentials of concentrations of all species equal zero, which necessarily describes the trailing solvent plateau. Moreover, it is possible to define the change explicitly. Substitution of dm_T from eqn (2.32b) into eqn (2.33) gives, on division by dm_S (which is permitted since $dm_S\neq 0$ in this region),

$$u=\frac{v_S+v_T m^2 Y(m_S)^{m-1}}{1+m^2Y(m_S)^{m-1}}. \tag{2.35}$$

For selected values of the constants v_S, v_T, Y, and m, it is possible to employ a range of values of m_S and hence calculate the corresponding values of u by means of eqn (2.35). Equation (2.32b) provides the corresponding values of m_T and eqn (2.34) the relevant values of $\bar{c}_L$ (if required). It also follows from eqn (2.35) that the limits of the boundary, viz. dm_S and $dm_T=0$, are at the points of discontinuity $v_S t$ and $v_T t$. This is an example of a *spread reaction boundary*, across which the concentrations of two species change, while equilibrium between the species is maintained at all points. The various forms which this type of boundary may assume for different values of m will be discussed fully in Chapter 3, where the behaviour of polymerizing systems is examined in greater detail.

The solution for the descending side is now complete. There are two constituents and hence two boundaries and three plateau regions (Nichol, Ogston, and Rescigno 1970). One spread boundary exists between $v_S t$ and $v_T t$, and a sharp boundary occurs at $v_A t$, the plateau between them containing unbound ligand in both forms, S and T, the concentrations of which are equal to those in the original equilibrium mixture.

(c) *A partial solution for the ascending side*

For the conjugate ascending side of the profile, directions are taken as positive from solution towards solvent, so that species are labelled as before. Possible alternative conventions will be discussed in Chapter 4. Equations (2.30), (2.31), and (2.32a, b) continue to apply, since they describe conservation of mass and the law of mass action.

(i) *The plateau regions of solvent and of solution of original composition.* When $u < v_S$ or $u > v_A$ it is obvious that $(u - v_A)$, $(u - v_S)$, and $(u - v_T) \neq 0$, whereupon eqns (2.30) and (2.31) are satisfied only by placing differentials of all species concentrations equal to zero. From the design of the experiment, this clearly implies that the solvent plateau exists when $u > v_A$, and the original solution plateau exists at least when $u < v_S$.

(ii) *The position* $x = v_A t$. When $u = v_A$, then $(u - v_A) = 0$, and eqn (2.31) reverts to eqn (2.33). Furthermore, since $(u - v_S)$ and $(u - v_T)$ are both greater than zero, it follows that $\mathrm{d}m_S = \mathrm{d}m_T = 0$ is a unique solution to eqn (2.33). In short, $m_S = m_T = 0$ when $u \geqq v_A$. However, eqn (2.30) permits $\mathrm{d}\bar{c}_P$ to be non-zero, and in the absence of S and T this would mean that the concentrations of A and C change in accordance with eqn (2.32a), this situation being illustrated in Fig. 2.3, ascending side. The alternative possibility that $\mathrm{d}\bar{c}_P = 0$ when $u = v_A$ is excluded by what follows.

(iii) *The region* $v_S < u < v_A$. At some position(s) in this region m_S and m_T must change from their original values to zero (i.e. proceeding from solution to solvent, $\mathrm{d}m_S$ and $\mathrm{d}m_T < 0$, according to eqn (2.32b)). It is not possible for $\mathrm{d}m_A$ and $\mathrm{d}m_C$ to be also negative, since eqn (2.30) insists that $\mathrm{d}\bar{c}_P = 0$ when, as in this region, $u \neq v_A$. Indeed, for the latter restriction to hold when $\mathrm{d}m_S$ and $\mathrm{d}m_T < 0$, $\mathrm{d}m_A$ and $\mathrm{d}m_C$ must be positive and the differentials of the concentrations of complexes negative. It follows that m_A and m_C must change at $x = v_A t$ to permit the required change in the region under discussion. Suppose that the concentrations of A and C increase at $v_A t$ from zero to m'_A and m'_C. According to eqn (2.30) these values are maintained in a plateau region provided $m_S = m_T = 0$: this plateau is terminated by a boundary, wherein, as we have seen, m_A and m_C increase while m_S and m_T decrease. In addition, since $d\bar{c}_P = 0$, the concentrations of all $AS_i T_j$ and $CS_k T_l$ must decrease (together with S and T) from their original solution values to zero.

An important conclusion follows the finding that the *constituent* concentration of acceptor, $\bar{c}_P$, does not change across this reaction boundary. It is that the concentration of acceptor in the leading plateau on the ascending side, where acceptor exists alone, is equal to its *total* concentration in the original solution, which has been determined by the investigator. This situation is attained only in the event that condition (2.27) holds, and thus provides an internal check of its validity. Such confirmation is important to the experimenter because in no other circumstance may the descending side be used to obtain binding results in the manner described above.

The complete solution of the ascending side would require specification of the form and position of the reaction boundary which must occur in the region $v_S < u < v_A$. This is difficult for the general case where the nature of the complexes $AS_i T_j$ and $CS_k T_l$ remain unspecified. Let us clarify the matter with a specific and simpler system.

(d) *A complete solution for the ascending side for the system* $A + S \rightleftharpoons AS$, *with* $v_S < v_A = v_{AS}$

In what follows, reference may be made to the left hand side of Fig. 2.4, which presents the ascending side (adopting the convention that directions are positive from solution to solvent as before). The conjugate descending side is included for comparison; but since the complete solution for this side has already been given above, we need only comment that the reaction boundary between $v_S t$ and $v_T t$ in Fig. 2.3, descending side, is replaced by a sharp boundary at $v_S t$ in Fig. 2.4.

By analogy with eqn (2.9) we may write expressions for the constituent concentrations $\bar{m}_A$ and $\bar{m}_S$ (on a molar scale) and the corresponding expressions for the constituent velocities, $\bar{v}_A$ and $\bar{v}_S$, as

$$\bar{m}_A = m_A + m_{AS}; \quad \bar{v}_A \bar{m}_A = v_A m_A + v_A m_{AS}; \tag{2.36a}$$

$$\bar{m}_S = m_S + m_{AS}; \quad \bar{v}_S \bar{m}_S = v_S m_S + v_A m_{AS}. \tag{2.36b}$$

These equations are consistent with the more general expressions (eqns (1.9),

(1.10), and (2.26a, b)), as may be seen by placing $m_C=0$, $m_T=0$, $p=1$, $\bar{c}_P/M_A=\bar{m}_A$, and $\bar{c}_L/M_S=\bar{m}_S$ in the general equations. Combination of eqns (2.36a, b) with the appropriate continuity equations, entirely analogous to eqns (2.11a, b), gives

$$(u-v_A)\,dm_A+(u-v_A)\,dm_{AS}=0, \tag{2.37a}$$

$$(u-v_S)\,dm_S+(u-v_A)\,dm_{AS}=0. \tag{2.37b}$$

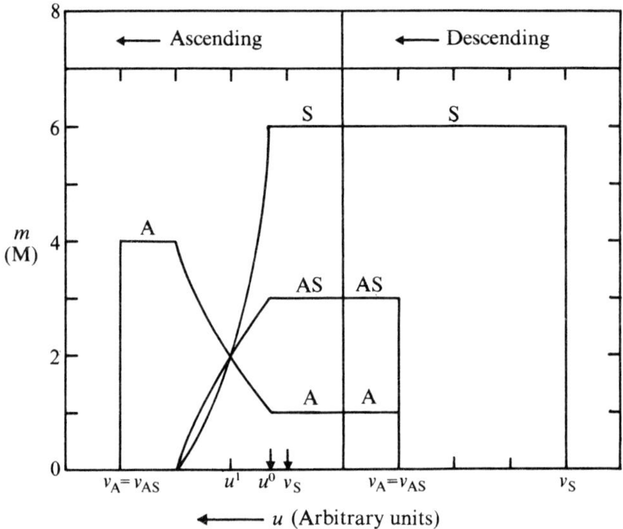

FIG. 2.4. Numerical example of a plot of concentration against the normalized distance parameter, u, for a system $A+S \rightleftharpoons AS$, with $v_S<v_A=v_{AS}$. Distances are positive in the direction solvent→solution (descending side) and solution→solvent (ascending side). Magnitudes of parameters are as follows: $\bar{m}_A=4$ M, $\bar{m}_S=9$ M, $K_A=0{\cdot}5\ \mathrm{M}^{-1}$, $v_A=v_{AS}=4$, $v_S=1$. (Adapted from Gilbert and Jenkins (1959) and Nichol and Ogston (1965*a*).)

The association equilibrium constant is given by

$$K=\frac{m_{AS}}{m_A m_S} \quad \text{or} \quad dm_{AS}=Km_A\,dm_S+Km_S\,dm_A. \tag{2.38}$$

When $u>v_A$ or $u<v_S$, we necessarily describe the solvent and solution plateaux, respectively, as before. On equating u and v_S eqn (2.37b) requires that $dm_{AS}=0$ (since $(u-v_A)\neq 0$), eqn (2.37a) that $dm_A=0$, and eqn (2.38) that $dm_S=0$. Consequently, the original solution plateau is observed when $u\leq v_S$. At the position $ut=v_A t$ eqn (2.37b) insists that $dm_S=0$ and, therefore, that

$m_S=0$: non-zero values of dm_A are permitted, and from above it may be concluded that m_A increases from zero to a plateau value given by the total concentration of A in the original solution.

The remaining region to examine is $v_S<u<v_A$. In this region $(u-v_A)\neq 0$, in which case eqn (2.37a) may be divided by this term to give

$$dm_A+dm_{AS}=d(m_A+m_{AS})=d\bar{m}_A=0. \tag{2.39}$$

Combination of eqn (2.39) with eqn (2.38) shows that

$$(1+Km_S)\ dm_A+Km_A\ dm_S=0. \tag{2.40}$$

Since $m_A\neq 0$ when $u<v_A$, *either* $dm_A=dm_S=0$, *or* the differentials are of opposite signs: the first possibility describes a plateau of A alone and the second a reaction boundary. More information may be obtained by combining the second continuity expression (eqn (2.37b)) with eqn (2.38) to give

$$\{(u-v_S)+(u-v_A)Km_A\}\ dm_S+(u-v_A)Km_S\ dm_A=0. \tag{2.41}$$

We recall that $m_S=0$ when $u\geqq v_A$, and if it continues to equal zero when $u<v_A$, then eqn (2.41) becomes

$$\{(u-v_S)+(u-v_A)Km_A\}\ dm_S=0, \tag{2.42a}$$

which shows that $dm_S=0$ or that

$$(u-v_S)+(u-v_A)Km_A=0. \tag{2.42b}$$

Thus a plateau of A alone at its total concentration in the original solution continues when $u<v_A$ until a point of singularity is reached where m_S begins to change. Representing this point by $u't$, we note that eqn (2.42b) pertains to a region including this point, whereupon rearrangement gives

$$u'=\frac{v_S+v_A KR}{1+KR}, \tag{2.43}$$

in which R is the particular value of m_A at u': this, we have seen, is necessarily the total concentration of A in the original solution.

Substitution of dm_A from eqn (2.40) into eqn (2.41), recalling that $dm_S\neq 0$, and rearrangement yields

$$u=\frac{v_S+v_S Km_S+v_A Km_A}{1+Km_S+Km_A}, \tag{2.44}$$

from which the point of singularity, $u^{\circ}t$, at which the boundary terminates in the original solution plateau is obtained by placing m_A and m_S equal to their values in the original solution plateau. Eqn (2.43) follows from eqn (2.44) on substituting the relevant values of m_S (=0) and m_A (=R) pertaining at $u't$.

We now wish to write explicit expressions for m_A, m_S, and m_{AS} as a function of u and the constants relevant to the system, since this would permit construction of the complete ascending pattern from known parameters. Equation (2.40) may be written as

$$\mathrm{d}\{(1+Km_S)m_A\}=0 \tag{2.45}$$

and integrated to the boundary condition of the plateau to give (together with eqn (2.36a) and the definition of R)

$$(1+Km_S)m_A=R. \tag{2.46}$$

Combination of this expression with eqn (2.44) yields

$$m_A=\left\{\frac{R(v_S-u)}{K(u-v_A)}\right\}^{1/2}, \tag{2.47a}$$

$$m_S=\left\{\frac{R(u-v_A)}{K(v_S-u)}\right\}^{1/2}-\frac{1}{K}, \tag{2.47b}$$

$$m_{AS}=R-\left\{\frac{R(v_S-u)}{K(u-v_A)}\right\}^{1/2}. \tag{2.47c}$$

Figure 2.4 presents a numerical example pertaining to a system characterized by the following parameters: $\bar{m}_A=4\,\mathrm{M}$, $\bar{m}_S=9\,\mathrm{M}$, $K=0{\cdot}5\,\mathrm{M}^{-1}$, $v_A=v_C=4$, $v_S=1$. u' was calculated from eqn (2.43), u^o from eqn (2.44), and the concentration gradients from eqns (2.47a, b, c) by simple arithmetic. The reaction boundary situated between $u't$ and u^ot could also have been plotted in terms of constituent concentrations as a function of u. On adding eqns (2.47a) and (2.47c) we find that $\bar{m}_A=R$, which stresses the fact that there is no change in this constituent concentration across the boundary. Addition of eqns (2.47b) and (2.47c) gives

$$\bar{m}_S=\left(R-\frac{1}{K}\right)+\left(\frac{R}{K}\right)^{1/2}\left\{\frac{(u-v_A)^{1/2}}{(v_S-u)^{1/2}}-\frac{(v_S-u)^{1/2}}{(u-v_A)^{1/2}}\right\}, \tag{2.48}$$

which shows that the concentration of ligand does change across the boundary between the positions $u't$ and u^ot.

Suppose, now, in relation to Fig. 2.4, that a method such as refractometry was employed to record the total concentrations of all species as a function of a distance parameter at a fixed time. The experimental pattern would exhibit a sharp boundary on the ascending side at the position $v_A\,t$, and a corresponding sharp boundary on the descending side, also at $v_A\,t$: for example, in gel chromatography the elution volumes would be identical after subtracting the loading volume from that observed on the descending side. However, the ascending and descending sides are non-enantiographic in two respects: (i) the magnitude of the boundary at $v_A t$ differs, since it corresponds to the appearance of the constituent concentration of A in the former and to the quantity $\bar{c}_P + M_S(\bar{m}_S - m_S)$ in the latter; (ii) a sharp boundary is observed at $v_S t$ on the descending side, but its counterpart on the ascending side is spread due to chemical reaction, is situated in a region between v_S and v_A, and is of different magnitude. The behaviour may be contrasted with that shown in Fig. 2.1, where enantiography of conjugate sides is evident for a mixture of non-interacting solutes. The present observation that a *rapid chemical reaction leads to migration patterns in which conjugate sides are non-enantiographic* will be shown later to be a general phenomenon. This is one reason for techniques such as frontal gel chromatography and moving boundary electrophoresis, which allow comparison of ascending and descending sides, being preferred to zonal methods for the study of interacting systems.

2.6. A practical example of the determination of binding results by frontal gel chromatography

The multiple binding of methyl orange (the ligand) and undefatted bovine serum albumin (the acceptor) in phosphate buffer, pH $5\cdot6_8$, ionic strength 0·11, has been studied by Klotz et al. (1946) by the method of equilibrium dialysis at 4°. This system therefore provides a suitable model for illustrating the application of frontal analysis in gel chromatography as a means of determining binding results (Nichol *et al.* 1971). A $1\cdot0 \times 38$ cm column (30 ml bed volume) of Sephadex G-50 was equilibrated with the phosphate buffer at 4°, and 60 ml ($=V_l$) of a solution containing 0·016 g/litre of methyl orange and 0·26 g/litre of bovine serum albumin (prepared by weight) was introduced onto the top of the column to form a sharp initial boundary between buffer and solution. When all of the solution had entered the column, elution was continued with the phosphate buffer, a flow rate of approximately 20 ml/h being maintained throughout. Fractions of approximately 2 ml (the actual volume of each being determined accurately by weight) were collected and then monitored spectrophotometrically at both 280 nm and 495 nm, wavelengths at which the absorption spectrum of methyl orange exhibits maxima: a molar extinction coefficient of $1\cdot8 \times 10^{+4}$ at 495 nm suffices to describe the absorbancy of methyl orange solutions, provided the concentration is less than 10^{-4} M. The result of the experiment is shown in Fig. 2.5, in which the abscissa axis of Fig. 2.4 (in terms of elution volumes) is reproduced to facilitate comparison with the theoretical result.

From the ascending side of Fig. 2.5, it is evident that a boundary corresponding to bovine serum albumin emerged from the column with an elution volume of 14 ml, the void volume of the column. Then followed a plateau region of protein alone, the absence of methyl orange being signified by the lack of light absorption at 495 nm. The concentration of protein in this region, 0·27 g/litre, is (within experimental error) the total protein concentration in the original solution (i.e. $\bar{c}_P = M_A R$). This observation shows not only that no irreversible binding of bovine serum albumin to the gel phase has occurred, but

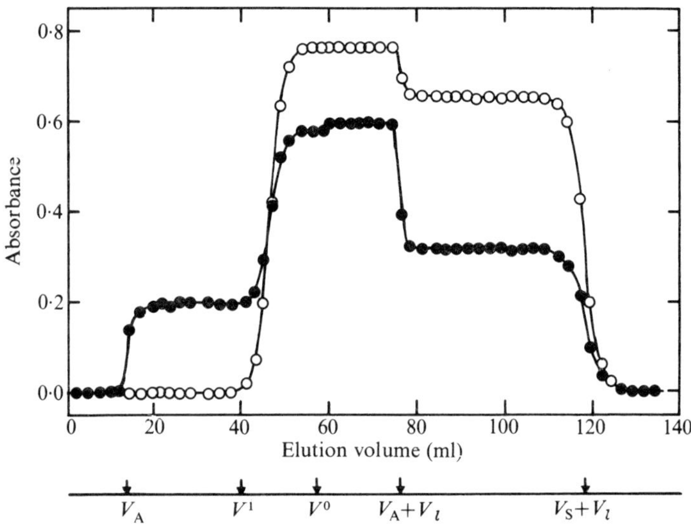

FIG. 2.5. An elution profile obtained in frontal gel chromatography of a mixture of methyl orange (0·016 mg/ml) and bovine serum albumin (0·26 mg/ml) on Sephadex G-50, the experiment being performed at 4° in 0·11 I phosphate buffer, pH 5·68. Open circles refer to absorbance readings at 495 nm, and filled circles to readings at 280 nm. The abscissa axis of Fig. 2.4 is included to facilitate comparison of the elution profile with that predicted theoretically. (Adapted from Nichol *et al.* (1971).)

also that all entities except uncomplexed methyl orange are excluded from the stationary phase (condition (2.27), or a simplified form applies). The protein plateau is terminated by a spread reaction boundary, situated between V' and V^0, across which concentrations of material absorbing at 495 and 280 nm change simultaneously. At V^0 the reaction is terminated by the plateau region of original composition, where the free and complexed forms of methyl orange contribute to the absorbancies at both wavelengths. On the descending side, the decrease in absorbancies at both wavelengths which terminates the latter plateau reflects the fall to zero of concentrations of unbound protein and of all complexes. This boundary is characterized by an effluent volume of 74 ml, which on subtraction of the 60-ml loading volume gives 14 ml, a value in excellent agreement with V_A. A plateau region of methyl orange alone follows, the ratio of absorbancies at the two wavelengths being consistent with this interpretation. It is from this plateau region that the equilibrium concentration of unbound methyl orange ($\bar{c}$) may be determined as 0·0119 g/litre (an average of twelve readings in this plateau region). A boundary with an elution volume of 54 ml (114–60) separates the plateaux of methyl orange and solvent. The value of 54 ml is in excellent agreement with that obtained when methyl orange was subjected to frontal analysis under the same conditions in a control experiment performed with a comparable concentration of dye. The large elution volume, corresponding to a distribution coefficient (Gelotte 1960) of 2·6, indicates that interactions between the dye and the Sephadex are not restricted to liquid–liquid partition between mobile and stationary phases. This factor in no way alters any conclusions drawn from the elution profile, provided it can be shown that interactions between solute and stationary phase (eqn (2.15)) are reversible, irrespective of their nature. In a series of control experiments performed with methyl orange alone it was found that the concentration in the plateau region was consistently 10 per cent lower than the applied concentration, indicating this degree of irreversible adsorption of the dye (Nichol *et al.* 1971). Such an observation highlights the advisability of performing control experiments with ligand and acceptor separately before the interpretation of results obtained with mixtures is attempted. Even in instances where irreversible binding is detected, it is possible, with some reservations, to estimate the value of r from eqn (1.2) by employing the observed value of $\bar{c}$ together with values of $\bar{c}_L$ and $\bar{c}_P$, corrected for any effects of irreversible adsorption. After correction of $\bar{c}_L$ for 10 per cent irreversible adsorption, the experiment shown in Fig. 2.5 yielded values of $0{\cdot}96 \times 10^{-2}$ and $1{\cdot}19 \times 10^{-2}$ g/litre for r and $\bar{c}$, respectively.

It is, of course, possible to repeat the above type of experiment with different constituent concentrations of ligand ($\bar{c}_L$) and thus obtain a series of (r, $\bar{c}$) points referring to a fixed value of $\bar{c}_P$. Nichol, Jackson, and Smith (1971) found that such results plotted in double-reciprocal form gave within experimental error a straight line, the slope and intercept of which gave, according to eqn (1.13), values of p (22 ± 5) and K_A ($2{\cdot}0 \pm 1{\cdot}0 \times 10^3\ \mathrm{M}^{-1}$) in good agreement with those reported by Klotz *et al.* (1946), who studied the same system by equilibrium dialysis. There is some justification for assuming that $\bar{c} = M_S\, m_S$ (i.e. that the ligand does not polymerize), since Beer's law is obeyed in the concentration range investigated ($<10^{-4}$ M). However, the acceptor, bovine serum albumin, comprises 85 per cent monomer and 15 per cent higher polymers (mainly dimer) which are not in rapid equilibrium (Squire, Moser, and O'Konski 1968). It follows from eqn (1.11) that eqn (1.13) may still be used to describe the binding results in the event that $K_A = K_C$ and $q = np$. Moreover, it is possible to show that within experimental error the results fail to distinguish between the latter model (equivalent binding of ligand to monomer and dimer) and an extreme model in which dimer does not bind ligand at all. Thus, on setting q and/or $K_C = 0$ in eqn (1.35) we obtain on rearrangement

$$\frac{1}{r} = \frac{\bar{c}_P}{pM_S K_A \bar{m}_A} \cdot \frac{1}{m_S} + \frac{\bar{c}_P}{pM_S \bar{m}_A}. \tag{2.49}$$

Since the ratio $\bar{c}_P/\bar{m}_A$ is constant, eqn (2.49) shows that a plot of $1/r$ vs. $1/m_S$ would still be linear in the event that only monomer were to bind ligand: this situation arises because the monomeric and dimeric bovine serum albumin are not in rapidly established equilibrium. The observed intercept of the linear double-reciprocal plot would be given by the second term on the right of eqn (2.49), leading to a mean value of 26 for p, which clearly lies within the range $p = 22 \pm 5$ found on the basis of eqn (1.13).

It may be noted that errors in binding results (which lead to some ambiguity in their interpretation) arise in the method of frontal analysis under discussion solely through uncertainty in concentration estimation. The method does not depend on an accurate knowledge of the elution volume axis, and in this respect is superior to the zonal method of Hummel and Dreyer (1962), which not only requires that condition (2.27) be fulfilled, but also requires an area determination involving an accurate assessment of both concentration and volume axes. However, in contrast with the Hummel and Dreyer method, that of frontal analysis requires relatively large amounts of both ligand and acceptor. The question arises as to what is the minimum volume of original solution that must be applied to the column in order to ensure a plateau region of the same composition in the elution profile. Reference to Figs. 2.4 or 2.5 shows that the extent of the required plateau is $(V_A + V_l) - V^o$, although in practice it will be less than this because of diffusional spreading of the boundaries. Since $V_S > V^o$, a positive value for the smaller difference $(V_A + V_l) - V_S$ would ensure the required plateau. This condition may be written

$$V_l > V_S - V_A, \tag{2.50}$$

which indicates that the applied volume be greater than the difference between the elution volumes of the reactants (found in separate experiments with the same column). At first sight, condition (2.50) suggests that use of a gel such that V_S approaches V_A would permit application of a small volume of mixture. However, the extent of the plateau containing ligand alone is also given by $(V_S - V_A)$, and thus opposing factors govern the selection of gel and of the value of V_l.

2.7. Conclusion

The formation of continuity equations written in terms of constituent concentrations and velocities, and expressing conservation of mass (neglecting diffusional flows) has pointed the way to the description of migration boundaries obtained with systems undergoing rapid chemical reequilibration. The particular example presented, involving a relative order of magnitudes of species velocities given by condition (2.27), has led to a result which may be useful for obtaining binding results with acceptor–ligand systems. Indeed, Nichol and Winzor (1964) have employed this method to determine the equilibrium constant describing the electrostatic interaction between ovalbumin and lysozyme, while its application to the bovine serum albumin–methyl-orange system described herein stresses its general applicability. (See also Cooper and Wood 1968). It was initially assumed in writing constituent quantities pertinent to the solution shown in Fig. 2.3 that binding was to equivalent and independent sites within each acceptor entity: this permitted eqns (2.28a, b) and (2.31) to be written in tractable form. In instances where the binding sites are non-equivalent it remains possible to derive eqns (2.30), (2.33), and (2.34), provided condition (2.27) continues to apply. Thus the basic conclusions expressed in Fig. 2.3 are valid regardless of whether binding sites are equivalent or not.

However, the present treatment was not intended solely to devise a method for obtaining binding results: after all, many other excellent methods are also available for this purpose. It was intended that the treatment should also introduce the concepts of reaction boundaries and of non-enantiography of conjugate ascending and descending sides separated by a plateau region of original composition, both of which are characteristic features of the migration of rapidly interacting systems. The approach has also raised some interesting questions, which remain to be answered. In particular, it was noted in the discussion of the descending side of Fig. 2.3 that the boundary separating the plateaux of solvent and of an equilibrium mixture of S and T ($m\text{S} \rightleftharpoons \text{T}$) was a reaction boundary described by eqn (2.35). This is an example of a boundary form obtained in the migration of a polymerizing system ($m > 1$), and, through the pioneer work of Gilbert, much greater detail is available on such boundaries, the subject of the next chapter.

3

POLYMERIZING SYSTEMS

3.1. The behaviour of a system comprising monomer in rapid equilibrium with a single higher polymer

THE simplest polymerizing system consists of monomer in rapidly established equilibrium with a single higher polymer ($nA \rightleftharpoons C$). Accordingly, we shall consider initially the characteristics of the migration patterns for these systems before proceeding to the more complex situation in which a series of polymers coexist.

(a) *Non-enantiography: spread and hypersharp boundaries*

Combination of eqns (2.6a), (2.6b), (2.7), and (2.8), which define the constituent concentration and velocity of the system $nA \rightleftharpoons C$ together with the definition of the association equilibrium constant (K) and the appropriate continuity expression, gives

$$(u + un(c_A)^{n-1}K - v_A - v_C n(c_A)^{n-1}K)\, dc_A = 0. \qquad (3.1)$$

Equation (3.1) has two solutions. The first is $dc_A = 0$ (and hence $dc_C = 0$), which describes the solvent plateau, wherein $c_A = c_C = 0$, and the original solution plateau, preserved throughout the experiment in velocity sedimentation, moving boundary electrophoresis and frontal gel chromatography, as described in Chapter 2. The second solution of eqn (3.1) may be written as

$$u = \frac{v_A + v_C n(c_A)^{n-1}K}{1 + n(c_A)^{n-1}K}, \qquad (3.2)$$

which is entirely analogous to eqn (2.35), wherein both concentrations and the equilibrium constant were expressed on a molar basis: the corresponding quantities in eqn (3.2) are expressed on the g per unit volume scale. Equation (3.2) describes the *single* boundary which separates the solvent and solution plateaux. Consequently, a single boundary is predicted on both ascending and descending sides for a single associating solute ($nA \rightleftharpoons C$), and in this respect the behaviour parallels that for a single, non-reacting solute (see Chapter 2). However, the similarity ceases with this observation; because, in contrast to the enantiographic sharp boundaries observed in conjugate limbs during migration of a non-reacting solute, those observed with an associating solute are neither enantiographic nor sharp. This may be seen by differentiating eqn (3.2) with respect to $\bar{c}$, the result, in reciprocal form, being

$$\frac{\mathrm{d}\bar{c}}{\mathrm{d}u}=\frac{\{1+n(c_{\mathrm{A}})^{n-1}K\}^{3}}{(v_{\mathrm{C}}-v_{\mathrm{A}})n(n-1)(c_{\mathrm{A}})^{n-2}K}. \tag{3.3}$$

With assigned values of n, K, v_{C}, and v_{A}, one may compute values of u (eqn (3.2)), of $\mathrm{d}\bar{c}/\mathrm{d}u$ (eqn (3.3)), and of $\bar{c}$ (eqns (2.6a) and (2.7)) for each of a selected series of values of c_{A}: a plot of $\mathrm{d}\bar{c}/\mathrm{d}u$ vs. u is analogous to a time-normalized *schlieren* pattern, while a plot of $\bar{c}$ vs. u is the corresponding integrated form. Clearly, the boundaries in either representation are reaction boundaries inasmuch as A and C coexist in equilibrium at all positions within them. (The restriction that only rapidly-established equilibria are being considered ensures, via eqn (2.8), that $\bar{v}$ and u are functions only of $\bar{c}$, and not of time.) What is the nature of the single reaction boundaries appearing on the ascending and descending sides? If the convention is adopted that directions are positive in the direction solvent→solution on *both* ascending and descending sides, it follows that $\mathrm{d}\bar{c}/\mathrm{d}x$ is positive on both sides. For the side on which $v_{\mathrm{C}}>v_{\mathrm{A}}$ (be it ascending or descending), it follows from eqn (3.3) that $\mathrm{d}\bar{c}/\mathrm{d}u$ is also positive, since $(v_{\mathrm{C}}-v_{\mathrm{A}})>0$: numerical examples, calculated by the manner suggested above, would show that the reaction boundary was spread, and positioned between the limits $v_{\mathrm{A}}t$ and $v_{\mathrm{C}}t$. On the conjugate side, whether it be descending or ascending, $v_{\mathrm{A}}>v_{\mathrm{C}}$ according to the adopted convention, and thus $\mathrm{d}\bar{c}/\mathrm{d}u<0$ while $\mathrm{d}\bar{c}/\mathrm{d}x>0$. This is the condition, proposed by Nichol and Ogston (1965*a*, 1967), for the formation of a *hypersharp boundary*. The origin of hypersharp boundaries and their description in terms of step-functions have been discussed by DeVault (1943), Makino and Rogers (1965*a*, *b*), and Nichol *et al.* (1970). It suffices in relation to the boundary under discussion to point out that a change in concentration of both A and C occurs at a position $\bar{u}t$, the change being from zero (solvent plateau) to the original solution plateau values: in practice the tendency of the boundary to hypersharpen due to chemical reaction would oppose the tendency of the boundary to spread by diffusion.

The velocity of the median bisector of the gradient in constituent concentration of solute may be defined (Nichol and Ogston 1965*a*) as

$$\bar{u}=\frac{\int u\,\mathrm{d}\bar{c}}{\int \mathrm{d}\bar{c}}=\frac{\int \mathrm{d}\,(\bar{v}\bar{c})}{\int \mathrm{d}\bar{c}}=\bar{v}, \tag{3.4}$$

where the limits of integration are taken from the solvent plateau to the original solution plateau: this median bisector, also termed the centroid (Longsworth 1943), may be obtained from either the ascending or the descending side. It is clear from eqn (2.6b) that $\bar{u}=\bar{v}$, so defined, is the weight-average velocity of the system. For the hypersharp boundary, $\bar{u}$ is, of course, a parameter that defines its velocity uniquely and solely. In the case of the spread boundary, the integrations across the entire boundary, indicated by eqn (3.4), must be used to calculate the weight-average velocity.

In summary, the migration pattern obtained with a rapidly interacting system of the type $n\mathrm{A} \rightleftharpoons \mathrm{C}$ ($n>1$, $v_\mathrm{A} \neq v_\mathrm{C}$) comprises on one side a hypersharp reaction boundary moving with the weight-average velocity of the system, and on the conjugate side a spread reaction boundary positioned between the limits $v_\mathrm{A}t$ and $v_\mathrm{C}t$: non-enantiographic patterns are thus observed. In velocity sedimentation, which only records the descending side, $v_\mathrm{C}>v_\mathrm{A}$ with the adopted convention (since larger solutes generally migrate faster), and thus a spread reaction boundary is expected; because of the sector-shaped cell and non-uniform field that pertain in ultracentrifugation, the weight-average sedimentation coefficient is not obtained from the rate of migration of the median bisector of the boundary (eqn (3.4)), but from the movement of the square root of the second moment of the entire schlieren pattern (Goldberg 1953, Schachman 1959). In moving boundary electrophoresis the same velocity relationship ($v_\mathrm{C}>v_\mathrm{A}$) may apply to the descending side, whereupon the boundary on that side would be spread, and its ascending counterpart hypersharp: an example of such behaviour is provided by the electrophoresis of β-lactoglobulin (Tombs 1957). However, since electrophoretic mobilities are functions of size, shape, and charge distribution, it is possible that with certain systems the reverse relationship ($v_\mathrm{C}<v_\mathrm{A}$) may pertain on the descending side, in which case the hypersharp boundary would form on that side, even though velocity sedimentation of the same system would have yielded a spread descending boundary.

(h) *Further description of the spread reaction boundary*

Although eqns (3.2) and (3.3) may be used to construct numerical examples of migration patterns for associating solutes, Gilbert (1955, 1959) formulated an important generalization by differentiating eqn (3.3) with respect to u and placing $\mathrm{d}^2\bar{c}/\mathrm{d}u^2=0$. This procedure leads to the conclusion that when $n>2$, a single minimum exists in a plot of $\mathrm{d}\bar{c}/\mathrm{d}u$ vs. u at a position $u_{\min}t$, given by

$$\frac{u_{\min}-v_\mathrm{A}}{v_\mathrm{C}-v_\mathrm{A}}=\frac{n-2}{3(n-1)}. \tag{3.5}$$

Thus, when $n=2$ (a monomer–dimer system), the only inflexion point is at the position of the monomer, and a single boundary is observed (Fig. 3.1(a)); but when $n>2$, a *bimodal* reaction boundary is observed (Fig. 3.1(b)). Equation

(3.2) may be rearranged to give $c_A = f(v_A, v_C, n, K)$, whereupon eqn (2.7) may be used to obtain an expression for c_C. Addition of c_A and c_C gives, according to eqn (2.6a),

$$\bar{c} = \left(\frac{1}{K}\right)^{1/(n-1)} \left[\left\{ \frac{v_A - u}{n(u - v_C)} \right\}^{1/(n-1)} + \left\{ \frac{v_A - u}{n(u - v_C)} \right\}^{n/(n-1)} \right]. \tag{3.6}$$

In a plot of $d\bar{c}/du$ vs. u, the area enclosed by the curve and the baseline between $u = v_A$ and $u = u_{min}$ corresponds to the area under the more slowly migrating portion of the bimodal reaction boundary, and, accordingly equals $\bar{c}$ at $u = u_{min}$: we shall term this concentration $\bar{c}_{min}$. We may therefore equate $\bar{c} = \bar{c}_{min}$ in eqn (3.6) provided we substitute $u = u_{min}$, which may be written as an explicit function of n, v_A, and v_C, by means of eqn (3.5). This combination gives

$$K = (\bar{c}_{min})^{1-n} \frac{(n-2)\{2(n^2-1)\}^{n-1}}{\{n(2n-1)\}^n}. \tag{3.7}$$

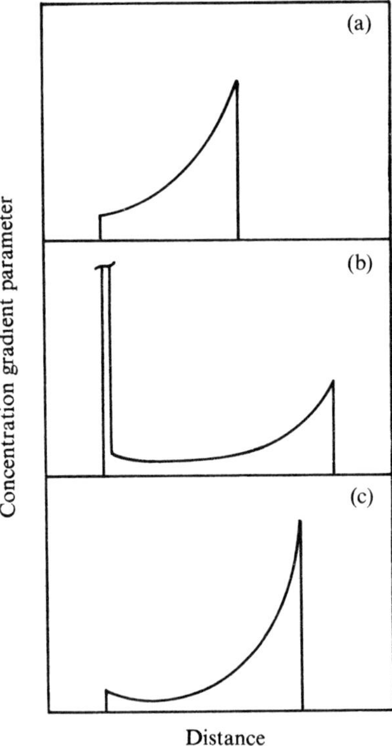

FIG. 3.1. Theoretical schlieren patterns calculated with arbitrary sets of parameters for (a) monomer–dimer, (b) monomer–trimer, and (c) monomer–dimer–trimer systems. ((b) and (c) adapted from Bethune and Kegeles (1961*c*).)

It is clear from eqn (3.7) that for a given system in a particular environment (i.e. for fixed values of K and n) $\bar{c}_{\min}$ is a constant: this conclusion, originally drawn by Gilbert (1955, 1959), is also implicit in eqn (3.5). This requirement leads to an interesting and characteristic concentration-dependence of the relative areas of the two portions of the bimodal reaction boundary. The largest value that u can attain is fixed by the total concentration of solute in the original solution plateau. If this total concentration is too low to give a value of u as large as $u_{\min}$, then only one schlieren peak is observed (which, nevertheless, represents gradients in concentration of both A and C). As the plateau concentration is increased, in separate experiments in the same environment, u can attain larger and larger values, until the situation is reached where u exceeds $u_{\min}$, and two peaks are observed. However, as the concentration is increased further, only the area of the faster peak increases, since $\bar{c}_{\min}$ is a constant. This type of behaviour has been observed in sedimentation velocity of mercuripapain (Smith, Kimmel, and Brown 1954), α-chymotrypsin (Massey, Harrington, and Hartley 1955; Nichol and Bethune 1963), β-lactoglobulin (Townend *et al.* 1960) and carboxypeptidase A (Bethune 1965*a*, *b*). In principle, determination of $\bar{c}_{\min}$ (expressed as g/unit volume) permits K to be determined using eqn (3.7), provided the value of n is known; but the method does not provide a means of estimating n. Moreover, the two portions of the bimodal reaction boundary (the fast and slow schlieren peaks) are never completely resolved in that the concentration (or refractive index) gradient never falls to zero between them: thus $\bar{c}_{\min}$ is difficult to determine accurately.

An alternative procedure for the determination of K involves the use of the weight-average velocity, $\bar{u}$, of the system. Regardless of whether $n=2$ or $n>2$, $\bar{u}$ may be determined from the rate of movement of the hypersharp boundary or from the conjugate spread boundary (be it uni- or bi-modal) by means of eqn (3.4). A single value of $\bar{u}$ is obtained for each value of $\bar{c}_0$ (total concentration in the original solution plateau) examined, a plot of $\bar{u}$ vs. $\bar{c}_0$ being a curve with an initially positive slope. Thus as $\bar{c}_0$ decreases, dissociation is favoured and $\bar{u}$ approaches v_A; while as $\bar{c}_0$ increases, increased association causes $\bar{u}$ to approach v_C. Indeed, the curve is described by an equation obtained by combining eqns (2.6a), (2.6b), (2.7a), and (3.4), viz.

$$K=\frac{(\bar{c}_0)^{1-n}(v_A-v_C)^{n-1}(v_A-\bar{u})}{(\bar{u}-v_C)^n}. \tag{3.8a}$$

Provided v_A and v_C may be determined, eqn (3.8a) may be used to calculate apparent K values from corresponding values of $\bar{u}$ and $\bar{c}_0$ for a selected range of values of n. The appropriate value of n (and hence K) may then be selected on the basis of the required insensitivity of K to variation in $\bar{c}_0$. Alternatively, the necessity of solving eqn (3.8a) for a series of values of n may be avoided by first combining eqns (2.6a) and (2.6b) to define c_A (eqn (3.8b)) pertaining to each experimental point ($\bar{c}_0$, $\bar{u}$). From the logarithmic form of the law of mass action for this system (eqn (3.8c)) it is evident

$$c_A = \frac{\bar{c}_0(\bar{u} - v_C)}{(v_A - v_C)}, \tag{3.8b}$$

$$\ln(\bar{c}_0 - c_A) = \ln K + n \ln c_A, \tag{3.8c}$$

that a plot of $\ln(\bar{c}_0 - c_A)$ vs. $\ln c_A$ yields both n and K.

The determination of v_A and v_C requires comment. Extrapolation of the $\bar{u}$ vs. $\bar{c}_0$ plot to infinite dilution permits an estimation of v_A, which might also be available from studies conducted in a medium where polymerization is inhibited and the monomer conformation unchanged. Gilbert and Gilbert (1960, 1961) have shown that the rate of movement of the slow peak of a bimodal reaction boundary does not correspond to v_A. Moreover, serious error is introduced if the rate of movement of the faster peak of the bimodal boundary (or an extrapolated value derived therefrom) is taken as a measure of v_C. The elegant theoretical examples used by Gilbert and Gilbert (1960, 1961), and their calculations relevant to the β-A and β-B lactoglobulin systems stress that neither peak represents monomer or polymer alone: there appears to be little further point in measuring the velocities of peaks in a bimodal reaction boundary, since they do not yield v_A, v_C, or $\bar{u}$. While estimates of v_A may be obtained by the extrapolation procedure described above, the determination of v_C is more difficult, because there is no defined position in the experimental record which corresponds to the position of C at a given time. One procedure, suggested by Nichol and Bethune (1963), involves rewriting eqn (3.5) as

$$\frac{(x_{\min} - x_A)}{(x_C - x_A)} = \frac{(n-2)}{3(n-1)}, \tag{3.9}$$

where $x_{\min}$, x_A, and x_C are the distances from the initial sharp boundary to the positions of the minimum in the schlieren pattern, the monomer (A), and polymer (C), respectively, after time t. $x_{\min}$ may be measured from the experimental pattern, x_A may be calculated from v_A, and x_C computed from eqn (3.9) for selected values of n, whereupon v_C (pertaining to each n) follows directly. Values of x_C may well be in the plateau region of original solution; certainly, they cannot be less than the distance from the initial sharp boundary to the position of the fast peak. For example, in velocity sedimentation of a 2 per cent solution of α-chymotrypsin in diethylbarbiturate buffer (pH $8{\cdot}4_5$, ionic strength 0·05, 20°), the observed sedimentation coefficients of the peaks comprising the bimodal reaction boundary were 2·7 S and 6·6 S (Nichol and Bethune 1963): from other studies (Massey *et al.* 1955, Neurath and Dreyer 1955) the value of s_A is 2·5 S, while that of s_C, computed with the aid of eqn (3.9), is 11·2 S (for $n=6$) or 10·7 S (for $n=8$). It is also of interest that the value of $\bar{c}_{min}$ was 3·6 g/litre in the veronal environment, compared with 1·8 g/litre in phosphate buffer, pH 7·9, ionic strength 0·03, and 20° (Massey *et al.* 1955), a change which can only reflect a variation in K or n (or both), inherent on the change of the ionic strength or of the isoelectric point of the α-chymotrypsin solutions (Anderson and Alberty 1948; Kubacki, Brown, and Laskowski 1949; Rao and Kegeles 1958). The determination of v_C from eqn (3.9) is limited by the accuracy with which x_{min} (the minimum position) may be determined, and is, of course, restricted to systems for which $n>2$ and for which bimodality is observed. In section 3 of this chapter, an alternative method of determining v_A and v_C (or their elution volume analogues) will be illustrated: it involves the determination of both weight- and z-average properties of the reaction boundary, and applies when $n \geqq 2$.

In Chapter 1, the importance was established of determining the stoichiometry of the polymerization and equilibrium constant for ligand as well as acceptor, these parameters being obtained from separate studies on the two solutes. Migration studies, together with the application of eqns (3.7) and (3.8), offer a potential means of achieving this purpose for simple systems of the type $nA \rightleftharpoons C$ (or $mS \rightleftharpoons T$). We have noted certain limitations involved in the use of $\bar{c}_{min}$ and $\bar{u}$ results, but one further difficulty merits attention. It has been assumed thus far that v_A and v_C are constants, whereas in practice the velocities of macromolecules generally exhibit a dependence on concentration. In modifying eqn (3.8) to account for this concentration dependence it is generally assumed (at least for proteins) that the dependence is identical for both v_A and v_C (Gilbert 1960, Gilbert and Gilbert 1961). However, there is little doubt that analysis of weight-average molecular weight (M_w) vs. $\bar{c}_o$ results is preferable for the determination of K and n, especially since M_A and M_C ($=nM_A$) are concentration-independent. The relevant expression for K is (Nichol *et al.* 1964)

$$K=\frac{(\bar{c}_o)^{1-n}(n-1)^{n-1}\{(M_w/M_A)-1\}}{\{n-(M_w/M_A)\}^n}. \tag{3.10}$$

Methods for determining weight-average molecular weights (for example, light scattering and equilibrium sedimentation) are not the subject of this work, and their use in studying interacting systems has been reviewed elsewhere (for example, Adams 1967, McKenzie and Nichol 1970). There is also little doubt, however, that in the initial characterization of (say) a polymerizing protein system, migration studies will be employed as a prelude to the more demanding equilibrium studies. It is for this reason that the present description of spread reaction boundaries has been given.

3.2. Monomer in rapidly established equilibrium with a series of higher polymers

When $n>2$, the existence of polymers of size intermediate between monomer and the largest polymer is likely, because of the improbability of multiple-bodied collisions in polymer formation. One possible pathway for the formation of polymers is shown in eqn (3.11), together with the association equilibrium constants for each step,

$$\begin{aligned}
&A_1 + A_1 \rightleftharpoons A_2; \qquad K_2' = m_{A_2}/m_{A_1}^2;\\
&A_2 + A_1 \rightleftharpoons A_3; \qquad K_3' = m_{A_3}/m_{A_2} m_{A_1};\\
&\quad \cdots \qquad\qquad\qquad \cdots\\
&A_{i-1} + A_1 \rightleftharpoons A_i; \qquad K_i' = m_{A_i}/m_{A_{i-1}} m_{A_1};\\
&\quad \cdots \qquad\qquad\qquad \cdots\\
&A_{n-1} + A_1 \rightleftharpoons A_n; \qquad K_n' = m_{A_n}/m_{A_{n-1}} m_{A_1};
\end{aligned} \tag{3.11}$$

where $i=1, 2, 3, \ldots, n$ represents the size of the polymer, with A_1 representing monomer and m_{A_i} molar concentrations as before. An alternative set of equilibrium constants is made available by noting (Klotz 1946) that

$$K_i = \prod_{i=1}^{i=n} K_i', \tag{3.12}$$

where $K_1 = 1$ is introduced for reasons of nomenclature. Thus, the equilibrium concentrations of polymers in eqn (3.11) may be expressed in terms of the equilibrium constants defined in eqn (3.12) as follows:

$$\begin{aligned}
m_{A_1} &= K_1 m_{A_1},\\
m_{A_2} &= K_2 m_{A_1}^2,\\
&\cdots\\
m_{A_i} &= K_i m_{A_1}^i,\\
&\cdots\\
m_{A_n} &= K_n m_{A_1}^n
\end{aligned} \tag{3.13}$$

It is assumed in this treatment that there exists only one form of each polymer A_i: Nichol *et al.* (1964) have presented expressions which consider a series of forms of each polymer (for example, linear, cyclic, etc.).

We may now write expressions for the constituent concentration, $\bar{c}$, and velocity, $\bar{v}$, of the single associating solute, which are extensions of eqns (2.6a) and (2.6b), respectively.

$$\bar{c}=\sum_{i=1}^{i=n} M_{A_i}m_{A_i}=M_{A_1}\sum_{i=1}^{i=n} iK_i m_{A_1}^i, \tag{3.14a}$$

$$\bar{v}\bar{c}=\sum_{i=1}^{i=n} M_{A_i}v_{A_i}m_{A_i}=M_{A_1}\sum_{i=1}^{i=n} iv_{A_i}K_i m_{A_1}^i, \tag{3.14b}$$

where M_{A_1} is the molecular weight of monomer. The continuity expression (eqn (2.8)) applies, regardless of the number of polymeric forms in rapid equilibrium with the monomer, and thus we may write, using eqns (3.14a, b),

$$u=\frac{\mathrm{d}(\bar{v}\bar{c})}{\mathrm{d}\bar{c}} \tag{3.15a}$$

$$=\frac{\sum\limits_{i=1}^{i=n} i^2 v_{A_i}K_i m_{A_1}^{i-1}}{\sum\limits_{i-1}^{i=n} i^2 K_i m_{A_1}^{i-1}}, \tag{3.15b}$$

$$=\frac{Q}{P}, \tag{3.15c}$$

where $M_{A_1}P=\mathrm{d}\bar{c}/\mathrm{d}m_{A_1}$ and $M_{A_1}Q=\mathrm{d}(\bar{v}\bar{c})/\mathrm{d}m_{A_1}$. This formulation is identical with that given by Gilbert (1959), except that $\bar{c}$ has been expressed on a weight basis (g/litre) rather than the base-molar scale used previously. Differentiation of eqn (3.15c) with respect to $\bar{c}$ gives

$$\frac{\mathrm{d}u}{\mathrm{d}\bar{c}}=\frac{\mathrm{d}m_{A_1}}{\mathrm{d}\bar{c}}\left\{\frac{PQ'-QP'}{P^2}\right\}, \tag{3.16}$$

Q' and P' being the first derivatives of Q and P, respectively, with respect to m_{A_1}. At a fixed time, t, we may write

$$Zt=\frac{t\,\mathrm{d}(\bar{c}/M_{A_1})}{\mathrm{d}x}=M_{A_1}^{-1}\frac{\mathrm{d}\bar{c}}{\mathrm{d}u}, \tag{3.17a}$$

where Z is the molar concentration-gradient parameter utilized by Gilbert

(1959). Combination of eqns (3.16) and (3.17a), recalling from eqn (3.15c) that $M_{A_1}P = d\bar{c}/dm_{A_1}$, gives

$$Zt = \frac{P^3}{(PQ' - QP')}. \tag{3.17b}$$

With selected values of i, K_i, M_{A_1}, and v_{A_i}, it is possible for a chosen series of values of m_{A_1} to calculate corresponding values of $\bar{c}$ (from eqn (3.14a)), u (from eqn (3.15b)), and Zt (from eqn (3.17b)). A plot of Zt vs. ut $(=x)$ corresponds to a schlieren pattern, while a plot of $\bar{c}$ vs. u corresponds to the time-normalized integrated form. Suppose, for example, one wished to compute the form of the migration pattern (neglecting diffusional flows) of a monomer–dimer–trimer–tetramer system: the relevant expressions for use in eqns (3.15c) and (3.17b) are

$$\begin{aligned}
P &= 1 + 4K_2 m_{A_1} + 9K_3 m_{A_1}^2 + 16K_4 m_{A_1}^3, \\
Q &= v_{A_1} + 4K_2 v_{A_2} m_{A_1} + 9K_3 v_{A_3} m_{A_1}^2 + 16K_4 v_{A_4} m_{A_1}^3, \\
P' &= 4K_2 + 18K_3 m_{A_1} + 48K_4 m_{A_1}^2, \\
Q' &= 4K_2 v_{A_2} + 18K_3 v_{A_3} m_{A_1} + 48K_4 v_{A_4} m_{A_1}^2.
\end{aligned} \tag{3.18}$$

Such numerical examples show that, provided $v_{A_{i+1}} > v_{A_i}$ (with directions positive from solvent→solution in both ascending and descending sides), a spread reaction boundary forms on the descending side, while the conjugate ascending side exhibits a hypersharp boundary ($d\bar{c}/dx > 0$ and $d\bar{c}/du < 0$) moving with a weight-average velocity, $\bar{u}$, given by eqns (3.4) and (3.14b). With reference to the spread reaction boundary, eqns (3.5)–(3.9) no longer apply, since a single equilibrium constant does not suffice to describe the system. It is possible to differentiate eqn (3.17b), but the complexity of the resulting equation even for relatively simple cases as described in eqn (3.18) renders examination for inflexion points impossible. Thus the precise number of inflexion points has not been established definitively in any case where three or more oligomers coexist. Numerical examples (cf. Figs. 1 and 3 of Gilbert (1959)) show that the reaction boundary may be distinctly bimodal, or appear as a single peak (schlieren pattern) with an asymmetric leading or trailing edge, the form depending essentially on the relative magnitudes of the K_i values selected. As the concentration of the largest polymer increases relative to the concentrations of the intermediate oligomers, a shoulder develops in the schlieren pattern, which develops bimodality on further increase of the proportion of largest polymer. This type of behaviour is also illustrated by the calculations of Bethune and Kegeles (1961*c*), who used Gilbert's equations to contrast the migration behaviour of a monomer–trimer system (cf.

Fig. 3.1(b)) with that of a monomer–dimer–trimer system containing appreciable amounts of dimer (cf. Fig. 3.1(c)). A more dramatic set of numerical examples performed by Bethune and Grillo (1967) has shown that with suitable selection of parameters a monomer–trimer–nonamer system yields a *trimodal* reaction boundary.

A plot of $\bar{u}$ (from eqn (3.4)) vs. $\bar{c}_0$ for such a system is a curve with an initially positive slope, as for the simpler $n\mathrm{A} \rightleftharpoons \mathrm{C}$ system. However, in the event that a series of polymers coexist, no single value of n when used in eqn (3.8) would lead to a concentration-independent value of K_i: failure of eqn (3.8) to describe the $\bar{u}-\bar{c}_0$ results thereby provides one indication of the presence of appreciable amounts of intermediate polymers. However, it is not suggested that curve-fitting procedures be used to evaluate the relevant K_i values from such data. Indeed, only the concentration dependence of accurately determined weight-average molecular weights should be used for this purpose. For further discussion on the application of the Steiner (1952) method of analysis to ideal systems, the reader is referred to Fujita (1962) and Nichol *et al.* (1964): extensions of this approach to non-ideal systems involved in either discrete or indefinite self-association equilibria are discussed by Adams (1967), who also gives the earlier references.

3.3. Interpretation of migration patterns obtained with α-chymotrypsin

Since the initial observation by Schwert (1949) that α-chymotrypsin undergoes reversible polymerization, this enzyme has been studied extensively as a model system. From these studies has emerged the concept of α-chymotrypsin as a monomer–dimer system in acetate–chloride buffer, pH 3·9, ionic strength 0·20 (Winzor and Scheraga 1964; Winzor, Loke, and Nichol 1967), while at low ionic strengths a monomer–hexamer equilibrium is believed to operate in the vicinity of the isoelectric region, viz. pH 8 (Gilbert 1955, 1959; Nichol and Bethune 1963; Ackers and Thompson 1965): at an intermediate pH value (6·2) the coexistence of monomers, dimers, and trimers has been postulated on the basis of molecular weight estimations (Rao and Kegeles 1958). In this section we shall employ gel chromatographic data to illustrate the use of migration patterns for characterizing the association under the first two sets of conditions, and thereby justify to some extent their consideration as monomer-single higher polymer systems. A suggested gel chromatographic procedure for characterizing equilibrium mixtures of several polymeric species (Ackers 1968, 1970) is also considered.

(a) *Association of α-chymotrypsin in phosphate, pH 7·9, ionic strength 0·03*

First we shall consider the migration behaviour of α-chymotrypsin under conditions where a monomer–hexamer equilibrium is believed to operate.

From eqn (3.5), a bimodal trailing boundary system is predicted, since $v_C > v_A$ on this side: as noted earlier, such patterns have been obtained in sedimentation velocity (Massey *et al.* 1955), and also in gel chromatography (Winzor and Scheraga 1963). An illustration of the migration pattern obtained by the latter technique is shown in Fig. (3.2), from which the non-enantiography of the advancing and trailing sides is clearly evident. In this connection it is noted that eqn (3.3) predicts a hypersharp boundary on the ascending side, whereas some spreading of this boundary has occurred, presumably due to the fact that the magnitude of the concentration-dependence of velocity is insufficient to outweigh the effects of diffusion: in keeping with such a postulate is the observation of an advancing boundary which is considerably sharper than that normally obtained with a non-associating solute having the same elution volume.

As noted by Nichol and Bethune (1963), one of the chief problems encountered in use of sedimentation velocity patterns to characterize such systems is the absence of a definitive value of v_C unless n is known, or vice versa, this dilemma being evident from eqn (3.9). However, the elution profile

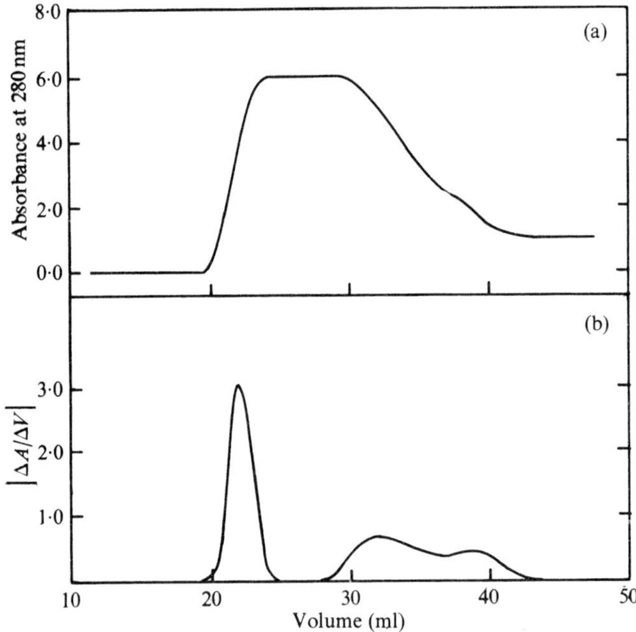

FIG. 3.2. Elution profile obtained in gel chromatography of α-chymotrypsin (3·8 mg/ml) on a 1·25 × 32 cm column of Sephadex G-100 equilibrated with 0·03 *I* phosphate, pH 7·9: the upper and lower patterns refer to the concentration and concentration gradient profiles, respectively, a value of 0·5 ml having been used for the volume increment in the latter. (Adapted from Winzor and Scheraga (1963).)

shown in Fig. 3.2 refers to gel chromatography on Sephadex G-100, and, consequently, oligomers for which $n > 5$ (i.e. molecular weight > 125 000) would be excluded from the gel phase, whereupon V_C, the chromatographic analogue of v_C, may be identified with V_0, the void volume of the column (11·9 ml in this instance). By performing a series of frontal runs with different plateau concentrations, V_A, the elution volume of monomer, may be obtained by extrapolation of $\bar{V}$, the median bisector of the sharp ascending boundary, to zero protein concentration: from Fig. 4 of Winzor and Scheraga (1963) a value of 25·0 ml for V_A is obtained. Substitution of these values in eqn (3.9), together with an estimate of 21·5 ml for V_{min} (inferred from Fig. 3.2) leads to a value of 6·0 for n (Ackers and Thompson 1965).

Having obtained a value for n, the next step in characterizing the system is to determine K, the association equilibrium constant, by means of eqn (3.7), in which K is now the only unknown, since $\bar{c}_{min}$ may be inferred from the size of the more slowly migrating part of the bimodal reaction boundary: Fig. 3.2 and Fig. 3 of Winzor and Scheraga (1963) suggest a value of approximately 1·4 g/litre for the latter quantity. By this means an estimate of $1{\cdot}5 \times 10^{-2}$ litre^5 g^{-5} is obtained for K. In this connection it is noted that the sedimentation velocity patterns of Massey *et al.* (1955) yield a slightly higher estimate of $\bar{c}_{min}$ (1·8 g/litre), from which an association equilibrium constant of $0{\cdot}4 \times 10^{-2}$ litre^5 g^{-5} is obtained. On conversion of these two values of K to molar quantities and thence to the corresponding standard free energy changes, estimates of −26·0 and −25·4 kcal are obtained for ΔG^0 from gel filtration and sedimentation velocity, respectively. Clearly the agreement between the two experimental methods is eminently satisfactory. But it is stressed again that the ultracentrifugal results were not amenable to unambiguous interpretation without prior knowledge of n, which we were able to obtain from the gel filtration pattern because of the insensitivity of V_C to the nature (within limits) of the higher polymer.

Despite the availability of a wide selection of gels with different permeation characteristics, it seems obvious that the above method of defining V_C, and hence n and K, has limited application. Furthermore, only in the event that $n > 2$ is a bimodal reaction boundary likely to be observed (eqn (3.5)), and so a more general procedure is required. Such a method, again employing gel chromatography, will now be illustrated.

(b) *Association of α-chymotrypsin in acetate-chloride, pH 3·86, ionic strength 0·2*

As mentioned in section 1(a) of this chapter, the weight-average velocity of a polymerizing solute ($n\text{A} \rightleftharpoons \text{C}$) may always be determined by applying eqn

(3.4) to either an ascending or descending migration pattern. In the event that the velocities of monomer and polymer, v_A and v_C, respectively, may also be determined, the values of n and K follow from eqn (3.8). Ackers and Thompson (1965) and Chiancone *et al.* (1968) have adopted this approach in gel chromatographic investigations of haemoglobin dissociation. Unfortunately, conditions cannot always be selected such that A and C can be obtained in pure forms. In experiments where both species coexist in appreciable amounts there is no defined position in the reaction boundary corresponding to either v_A or v_C, which cannot, therefore, be determined directly. α-chymotrypsin under this second set of conditions represents such a system where v_A, v_C, n, and K are all to be determined. The following procedure, written specifically in chromatographic terms but applicable in principle to all forms of frontal migration experiment, provides a means of estimating values of these four parameters (Winzor *et al.* 1967).

The weight- and z-average elution volumes, V_w and V_z, respectively, of an applied solution comprising an equilibrium mixture of monomer and single higher polymer are defined as

$$V_w = \frac{c_{o,A} V_A + c_{o,C} V_C}{c_{o,A} + c_{o,C}} = \frac{m_{o,A} V_A + n m_{o,C} V_C}{m_{o,A} + n m_{o,C}}, \tag{3.19a}$$

$$V_z = \frac{c_{o,A} V_A^2 + c_{o,C} V_C^2}{c_{o,A} V_A + c_{o,C} V_C} = \frac{m_{o,A} V_A^2 + n m_{o,C} V_C^2}{m_{o,A} V_A + n m_{o,C} V_C}. \tag{3.19b}$$

Subscripts o,A and o,C refer to concentrations of monomer and polymer, respectively, in the original solution plateau. Combination of eqns (3.19a) and (3.19b) leads to the expression (cf. Sophianopoulos and Van Holde 1964).

$$V_z = (V_A + V_C) - \left(\frac{V_A V_C}{V_w}\right). \tag{3.20}$$

Since V_w is determined readily there only remains the problem of estimating V_z from experimental profiles in order to define V_A and V_C from a plot of V_z vs. $1/V_w$ (eqn (3.20)).

The spread reaction boundary on the trailing side of a frontal gel filtration experiment may be characterized by the following weight- and z-average elution volumes, which apply to the general case of successive polymerizations (eqn (3.13)).

$$\bar{V} = \frac{\int_0^{\bar{c}_o} V \, d\bar{c}}{\int_0^{\bar{c}_o} d\bar{c}}, \tag{3.21a}$$

$$V_z^* = \frac{\int_0^{\bar{c}_0} V^2 \, \mathrm{d}\bar{c}}{\int_0^{\bar{c}_0} V \, \mathrm{d}\bar{c}} . \tag{3.21b}$$

As already proven (eqn (3.4)), $\bar{V}$ so defined is the weight-average elution volume of the applied solution (V_w), but the corresponding z-average quantities are not generally identical. However, a relationship between V_z, the z-average elution volume of the solution, and V_z^*, the z-average elution volume of the reaction boundary, may be found by combining eqns (3.15) and (3.21b), whereupon we obtain

$$V_z^* = \frac{\int_0^{\bar{c}_0} (Q/P)^2 \, \mathrm{d}\bar{c}}{\int_0^{\bar{c}_0} (Q/P) \, \mathrm{d}\bar{c}} \tag{3.22a}$$

or, with change of variable from $\bar{c}$ to m_{A_1}, the free monomer concentration (molar scale)

$$V_z^* = \frac{\int_0^{m_{0,A(1)}} (Q/P)^2 (\mathrm{d}\bar{c}/\mathrm{d}m_{A_1}) \, \mathrm{d}m_{A_1}}{\int_0^{m_{0,A(1)}} (Q/P)(\mathrm{d}\bar{c}/\mathrm{d}m_{A_1}) \, \mathrm{d}m_{A_1}} . \tag{3.22b}$$

Since

$$\frac{\mathrm{d}\bar{c}}{\mathrm{d}m_{A_1}} = \frac{\mathrm{d}}{\mathrm{d}m_{A_1}} \sum_{i=1}^{i=n} M_{A_1} i m_{A_i} = \sum_{i=1}^{i=n} M_{A_1} i^2 K_i (m_{A_1})^{i-1} = P M_{A_1} \tag{3.23a}$$

it follows that

$$V_z^* = \frac{\int_0^{m_{0,A(1)}} (Q^2/P) \, \mathrm{d}m_{A_1}}{\int_0^{m_{0,A(1)}} Q \, \mathrm{d}m_{A_1}} . \tag{3.23b}$$

We now apply this general equation to the specific case of a monomer-single higher polymer system, noting that $K_1 = 1$ and using subscripts A and C to denote monomer and polymer, respectively: eqn (3.24a) is obtained.

$$V_z^* = \frac{\int_0^{m_{0,A}} \{(V_A + n^2 V_C K_n m_A^{n-1})^2/(1 + n^2 K_n m_A^{n-1})\} \, \mathrm{d}m_A}{\int_0^{m_{0,A}} (V_A + n^2 V_C K_n m_A^{n-1}) \, \mathrm{d}m_A} . \tag{3.24a}$$

This may be rearranged further (Winzor *et al.* 1967) to give the expression

$$V_z = V_z^* + \frac{(V_A - V_C)^2}{V_A m_{0,A} + nV_C K_n m_{0,A}^n} \int_0^{m_{0,A}} \frac{n^2 K_n m_A^{n-1}}{1 + n^2 K_n m_A^{n-1}} \mathrm{d}m_A. \qquad (3.24b)$$

Replacement of K_n, the equilibrium constant expressed on a molar basis, by K, its weight-scale counterpart, and a change of variable from m_A to α_A, the weight-fraction of monomer, yields the expression

$$V_z = V_z^* + \frac{(V_A - V_C)^2 \alpha_{0,A} F}{V_w}, \qquad (3.25a)$$

where

$$F = \frac{1}{\alpha_{0,A}} \int_0^{\alpha_{0,A}} \frac{n(1-\alpha_{0,A})\alpha_A^{n-1}}{\alpha_{0,A}^n + n(1-\alpha_{0,A})\alpha_A^{n-1}} \mathrm{d}\alpha_A. \qquad (3.25b)$$

$\alpha_{0,A}$ may assume any value between 0 and 1, the same limits applying to the parameter F, which is, however, a function of both $\alpha_{0,A}$ and n: the dependence of F on $\alpha_{0,A}$ has been tabulated for values of n from 2 to 8 inclusive (Winzor *et al.* 1967). Numerical calculations have shown that the discrepancy between V_z and V_z^* is unlikely to exceed 6 per cent (Fig. 6 of above reference), whereupon V_z^*, the z-average elution volume of the reaction boundary, may be used as an initial estimate of V_z. A plot of this quantity against the reciprocal of V_w thus yields approximate values of V_A and V_C (eqn (3.20)). These estimates may then be used in eqn (3.25a) to give a second approximation to V_z for selected values of n, the required value of $\alpha_{0,A}$ being determined from eqn (3.19a). The relevant value of n is then obtained from the mass action law written in logarithmic form (eqn (3.8c)): only one set of values of $c_{0,A}$ as a function of $\bar{c}_0$ yields a straight line of the required slope (n) when log $c_{0,C}$ is plotted against log $c_{0,A}$, the value of K following from the ordinate intercept of the appropriate plot.

Figure 3.3(a) presents a plot of V_z^* vs. $1/V_w$ for α-chymotrypsin in 0·2 ionic strength acetate–chloride buffer, pH 3·86, these results referring to elution profiles in which $\bar{c}_0$, the plateau concentration varied from essentially zero to 3 mg/ml (Winzor and Scheraga 1964). Least squares calculations on these data yield values of 26·9 and 23·6 ml for the first approximations to V_A and V_C, respectively, while the plot of eqn (3.8c) (Fig. 3.3(b)) implicates dimer as the polymer in rapid association equilibrium with monomer. Application of the suggested reiterative procedure leads to estimates of V_z differing by only 0·06 to 0·10 ml from the corresponding V_z^* values. Initial and revised estimates of

V_z are thus identical within the limits of experimental accuracy of the elution volumes (±0·10 ml). On construction of a line with slope 2 through the point $(\log(\bar{c}_0 - c_{0,\mathrm{A}}), \log c_{0,\mathrm{A}})$, the mean of the experimental data, a value of 0·85 for the ordinate intercept (log K) is found, which corresponds to an association equilibrium constant of 7 litres/g. This value is in very good agreement with that of 4 litres/g inferred from weight-average molecular weight data on the same system (Winzor and Scheraga 1964).

Although this procedure avoids the necessity of obtaining estimates of oligomeric elution volumes from separate experiments, a factor barring its

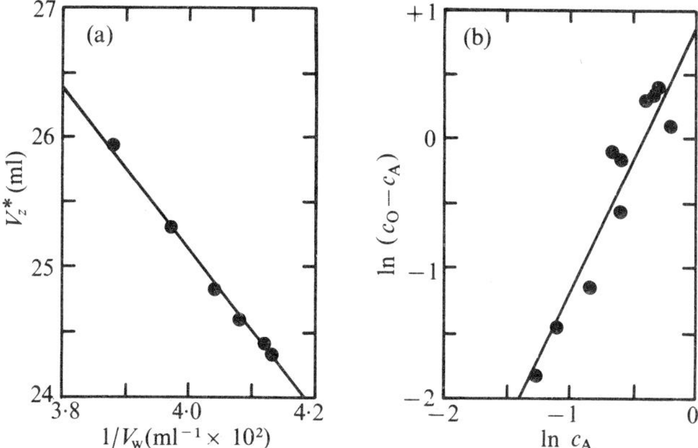

FIG. 3.3. Evaluation of V_A, V_C, n, and K for α-chymotrypsin in acetate–chloride buffer (I 0·20, pH 3·86): (a) the linear plot from which V_A and V_C were calculated (eqn (3.20)); (b) the plot of eqn (3.8c) with values of c_A calculated from V_w, V_A, and V_C. (Adapted from Winzor et al. (1967).)

application to *all* monomer-polymer systems is the assumed absence of intermediate polymers. In this connection it is noted that experimentally significant non conformity of an experimental system with the assumed coexistence of monomer and *single* higher polymer would be indicated by curvature of the V_z vs. $1/V_w$ plot (Sophianopoulos and Van Holde 1964). Evaluation of equilibrium constants for such systems represents a more difficult task, which will be considered in the next section on α-chymotrypsin at pH 6·2, conditions under which this enzyme is believed to exist in a series of polymeric forms (Rao and Kegeles 1958).

(c) *Association of α-chymotrypsin in phosphate buffer, pH* 6·2, *ionic strength* 0·20

Velocity sedimentation of α-chymotrypsin under these conditions yields a single, fairly symmetrical boundary, the plot of $\bar{s}$ vs. $\bar{c}_0$ exhibiting an initially positive slope (Schwert 1949, Rao and Kegeles 1958). At first sight this result seems to implicate a monomer–dimer equilibrium, but in a study of the concentration dependence of the weight-average molecular weight of α-chymotrypsin, values in excess of the dimeric value were obtained, and the results were accordingly interpreted in terms of the Steiner (1952) method of evaluating successive equilibrium constants: a monomer–dimer–trimer system was postulated (Rao and Kegeles 1958). The presence of more than one polymeric species may often be detected by, for example, non-linearity of the above two-species plot (Sophianopoulos and Van Holde 1964; Roark and Yphantis 1969; Harris, Kobes, Teller, and Rutter 1969) or by obvious non-conformity of the reaction boundaries in migration experiments with the expected shapes (Bethune and Grillo 1967). Evaluation of successive association equilibrium constants for such systems requires application of the Steiner (1952) procedure, or modifications thereof (Adams 1967). These procedures were written initially in terms of weight-average molecular weights, but Ackers (1968) has recently recommended analogous use of chromatographic data.

At any total concentration $\bar{c}_o$ of a series of polymers A, B, C, D, etc., in association equilibrium, the weight fraction of monomer, $\alpha_{o,A}$, may be obtained from knowledge of the weight-average value of any suitable parameter, X_w (molecular weight, sedimentation coefficient, elution volume), and the corresponding monomeric value, X_A, by means of eqn (3.26): X_A is found by extrapolating X_w to zero $\bar{c}_o$.

$$\ln \alpha_{o,A} = \int_0^{\bar{c}_o} \left(\frac{X_A}{X_w} - 1 \right) \mathrm{d} \ln \bar{c}. \tag{3.26}$$

From the definition of weight-average quantities it also follows that

$$X_w \bar{c}_o = X_A(\alpha_{o,A}\bar{c}_o) + X_B K_2'(\alpha_{o,A}\bar{c}_o)^2 + X_C K_2' K_3'(\alpha_{o,A}\bar{c}_o)^3 + \ldots, \tag{3.27}$$

where K_2', K_3', etc., are the association constants for the equilibrium between monomer and preceding oligomer. In the event that the weight-average property being measured is molecular weight, the successive equilibrium constants are the only parameters to be evaluated, since the oligomeric molecular weights follow directly from M_A. In gel chromatography, however, the elution volumes of B, C, D, etc., must also be regarded as unknown, whereupon the complexity of the problem is doubled. Although Ackers (1968) has suggested the use of gels with different porosities to increase the number of simultaneous equations that may be brought to bear on the solution of eqn (3.27), it is obvious that the fewer parameters there are to be evaluated, the greater are the chances of obtaining a unique solution. Hence, in our opinion, use of molecular weight data is preferred for characterization of systems comprising a series of polymers in association equilibrium unless conditions (for example, concentration range) preclude such measurements.

3.4. Differential migration techniques

Differential techniques, in which velocity differences rather than absolute migration rates are measured, provide the experimenter with more accurate comparisons of solutions whose weight-average properties are very similar: thus differential migration methods find application in the comparison of (a) solutions of the same solute differing only slightly in concentration, (b) solutions of very similar solutes at the same concentration, and (c) solutions of the same solute under different conditions (to study effect of pH, ligand addition, etc.). Two types of differential procedure have been devised for this purpose, viz. the layering technique (Hersh and Schachman 1955, Gilbert 1966*b*), and the method of Richards and Schachman (1957), in which the two solutions are placed in the separate cavities of a double-sectored cell: the former type of experiment will be discussed first.

(a) *Layering technique*

In this first type of differential migration experiment a boundary is formed between two solutions whose migration characteristics are to be compared. The differential migration rate of two solutions of the same substance differing only with respect to its concentration is thus measured by careful layering of the solution of lower concentration over that of higher concentration (for gravitational stability), and observing the migration of the boundary so formed (Hersh and Schachman 1955, Chiancone *et al.* 1968).

From considerations of mass conservation in such experiments it follows that*

$$\bar{v}\bar{c} + v_\Delta \Delta\bar{c} = (\bar{v} + \Delta\bar{v})(\bar{c} + \Delta\bar{c}) \tag{3.28}$$

where $\bar{v}$ is the constituent velocity corresponding to the constituent concentration $\bar{c}$ in the solution of lower concentration, $\Delta\bar{c}$ and $\Delta\bar{v}$ representing the respective changes in $\bar{c}$ and $\bar{v}$ across the differential boundary, the velocity of which is v_Δ. Equation (3.28) may be rearranged to give

$$v_\Delta = \frac{\Delta(\bar{v}\bar{c})}{\Delta\bar{c}}, \tag{3.29}$$

* For rigorous proof of eqn (3.28) the reader is referred to the derivation of eqn (5.52), for which it is apparent that v_Δ is the median bisector of the constituent concentration gradient in the region of the differential boundary.

from which it is evident (Chiancone *et al.* 1968, Gilbert and Gilbert 1968) that

$$\lim_{\Delta\bar{c}\to 0} v_\Delta = \frac{\mathrm{d}(\bar{v}\bar{c})}{\mathrm{d}\bar{c}} = u, \tag{3.30}$$

which is clearly the fundamental mass balance equation describing diffusion-free migration (eqn (2.8)). Ability to define u pertaining to a constituent concentration $\bar{c}$ appears to have two major applications, viz. the determination of equilibrium constants and of asymptotic migration patterns for self-interacting solutes. For a monomer-single higher polymer system the relevant expression for evaluating K is eqn (3.6), from which it is clear that values of v_A, v_C, and n are also required. In this connection, Chiancone *et al.* (1968) have also considered the stepwise formation of tetramer from monomer, a far more complicated situation in that four equilibrium constants and four species velocities must be incorporated into an expression analogous to eqn (3.6). Thus, not only may $(\bar{v}, \bar{c})$ results be used to evaluate equilibrium constants for polymerizing systems, but also $(u, \bar{c})$ data, the latter having the advantage that they may frequently be obtained over a far wider range of $\bar{c}$ values because of the differential method of measuring u.

The second suggested use of differential velocity experiments of this type is the determination of asymptotic migration patterns by measuring v_Δ in a series of runs with different values of $\bar{c}$. For purposes of illustration it is simplest to consider such profiles for an interaction which gives rise to linear concentration dependence of the constituent velocity, for example, of the form

$$\bar{v} = v^{\circ} - b\bar{c}, \tag{3.31}$$

where v° is the value of $\bar{v}$ at infinite dilution. Substitution of this relationship for $\bar{v}$ into eqn (3.30), and differentiation of the resulting expression, yields

$$u = v_\Delta = v^{\circ} - 2b\bar{c}. \tag{3.32}$$

Thus the dependence of u on $\bar{c}$ is double that of $\bar{v}$ on $\bar{c}$ for this simple case.

The experimental feasibility of obtaining diffusion-free migration patterns ($\bar{c}$ vs. u profiles), is illustrated in Fig. 3.4, which summarizes results obtained in velocity sedimentation of ovalbumin solutions. The open circles describe the dependence of s_{20} on concentration, which is approximately linear ($\bar{s} = s^{\circ} - b\bar{c}$), reflecting physical interactions between ovalbumin molecules: results of differential experiments with $\Delta\bar{c} = 0{\cdot}1$ g/dl, form the lower set of experimental points (closed circles). Least squares calculations on the two sets of results yield the relationships $s = 3{\cdot}4_5 - 0{\cdot}33\bar{c}$ and $s = 3{\cdot}3_9 - 0{\cdot}66\bar{c}$, the difference between the two ordinate intercepts being insignificant statistically. Thus the differential sedimentation coefficients have clearly described the dependence of u on concentration, whereupon the correct asymptotic profile would result.

Gilbert (1966*b*) has also elaborated on the possibility of employing the layering technique for a further purpose, viz. to compare the association constants of two polymerizing systems with identical V_A and V_C values by using a solution of one solute to elute a gel chromatographic column equilibrated with an identical concentration of the other solute. This technique seems eminently suited for evaluating the relative association constants of haemoglobin variants, or the various isoenzymes resulting from polymerization of parent A and A′ forms. Let us consider, for example, the result of performing such an experiment with two dimerizing systems for which $V_A = V_{A'} = 10{\cdot}0$ ml and $V_C = V_{C'} = 8{\cdot}5$ ml, the association constants for the

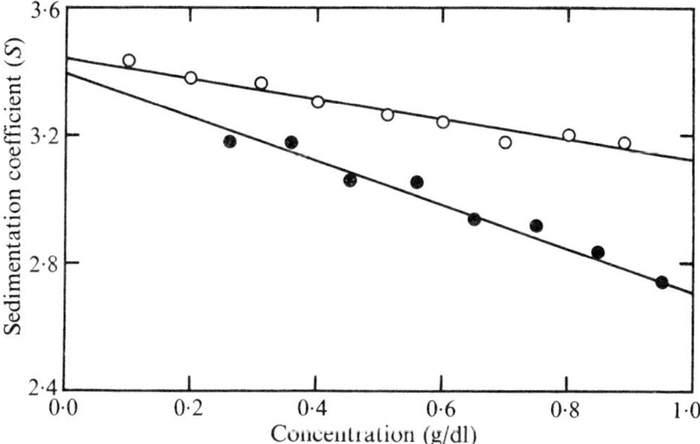

FIG. 3.4. Concentration dependence of the sedimentation behaviour of ovalbumin in 0·1 *I* phosphate, pH 6·8: O, sedimentation coefficients (s_{20}); ●, differential sedimentation coefficients (s_Δ) obtained with a concentration increment of 1 mg/ml.

equilibria 2A$\rightleftharpoons$C and 2A′$\rightleftharpoons$C′ being 24 and 12 litres/g, respectively (Gilbert and Gilbert 1968). Figure 3.5(a) represents the separate elution profiles for 0·08mg/ml solutions of the two solutes, median bisectors being used to denote the positions ($\bar{V}$) of the advancing (broken line) and trailing (solid line) profiles. Since A, the more associated solute, has been chosen arbitrarily as the solute with which the column was originally equilibrated, the observed pattern (sum of the two separate profiles) shows a dip in the region between $\bar{V}_A$ and $\bar{V}_{A'}$, but then rises again to $\bar{c}_0$ (0·08 mg/ml). Clearly, the magnitude of this dip is $\bar{c}_0(\bar{V}_{A'} - \bar{V}_A)$, from which $\bar{V}_{A'}$, the weight-average elution volume of A′ may be obtained provided $\bar{V}_A$ is known. In actual practice the observed concentration profile will generally exhibit both a dip and a hump at the junction of the two solutions because of the non-enantiography of ascending and descending sides for associating systems: the quantity $\bar{c}_0(\bar{V}_{A'}-\bar{V}_A)$ refers to the *net* increase or decrease. Figure 3.5(b), which presents the asymptotic profiles for the above system, illustrates this point.

As a method of estimating differences between solutes this procedure obviously reduces the experimental uncertainty inherent in inferences from separate experiments. Thus, while it is true that $\bar{V}_A$ and $\bar{V}_{A'}$ obtained from separate experiments could have been used to determine association constants for the dimerization of A and A′, respectively, the advantage of the differential experiment is that it provides a direct measure of the difference between c_A and $c_{A'}$, the respective concentrations of monomeric A and A′ present in the two

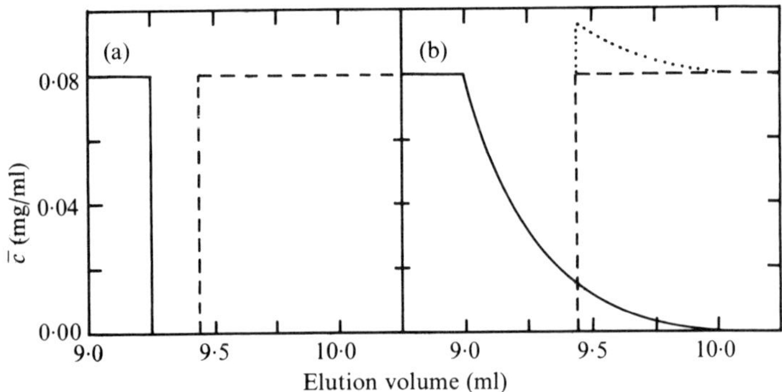

FIG. 3.5. Theoretical elution profiles calculated for a layering type differential experiment in which a 0·08 mg/ml solution of dimerizing solute A′ ($K = 12$ litres/g) is used to elute a column equilibrated with a 0·08 mg/ml solution of dimerizing solute A ($K = 24$ litres/g), the elution volumes of monomeric and dimeric species being 10·0 and 8·5 ml, respectively: (a) simplified profile in which only the median bisectors of boundaries are indicated; (b) the situation pertaining when the asymptotic shapes of profiles are taken into account.

solutions. This is evident from substitution of eqn (3.19a) for the weight-average elution volumes $\bar{V}_A$ and $\bar{V}_{A'}$, whereupon we obtain the expression

$$\bar{c}_0(\bar{V}_{A'}-\bar{V}_A)=(c_{A'}-c_A)(V_A-V_C).$$

The effect of environmental changes on the extent of association of a single solute has also been studied by differential gel chromatography (Baghurst, Nichol, Richards and Winzor, 1971).

(b) *Method of Richards and Schachman (1957)*

This differential method was devised principally for measuring differences between sedimentation coefficients of solute systems. The two solutions, containing identical macromolecular concentrations, are placed separately in the two sectors of a double-sectored cell, precautions being taken to ensure that the sectors are filled to exactly the same level. At the commencement of centrifugation the macromolecular concentrations are clearly identical at all levels in the two sectors, and so the Rayleigh pattern, which compares the concentrations at conjugate levels in the two solutions, comprises a series of straight fringes. If the rate of migration in two sectors is identical, the same Rayleigh pattern prevails throughout the experiment: despite the existence of boundaries in both sectors, the concentrations at conjugate levels are identical. With a faster migration rate in one sector this situation no longer pertains, with the result that concentration differences are observed as curved Rayleigh fringes. From the rate of change with time of the first moment of the pattern it is possible to calculate the quantity $\Delta\bar{s}/\bar{s}_{av}$, where $\Delta\bar{s}$ is difference between sedimentation coefficients pertaining to the two sectors, and $\bar{s}_{av}$ their mean*.

Interpretation of the ratio $\Delta\bar{s}/\bar{s}_{av}$ depends on the source of the difference between migration rates in the two solutions. If, for example, the two sectors contained identical concentrations of very similar proteins A and B, the experiment is an alternative to the layering experiment proposed by Gilbert (1966*b*), since it yields essentially $(\bar{s}_B - \bar{s}_A)$, which may therefore be used in precisely the same way as $(\bar{V}_B - \bar{V}_A)$ to define K for B relative to K for A. Secondly, since the two solutions are in separate compartments it is possible to investigate the effect of environmental factors on $\bar{s}$ by observing differential migration of identical concentrations of the same solute in different solvents. For example, from differential sedimentation of a non-self-interacting protein in buffer and deuterated buffer it is possible to estimate the partial specific volume of the solute (Martin, Cook, and Winkler 1956). Of more relevance in the present context is the effect of ligand addition to a self-interacting acceptor. Provided sufficient inert solute is added to the ligand-free solution of acceptor to counter the refractive index increment of added ligand, differential experiments of this type may be used to detect preferential binding of ligand to one form of acceptor. We illustrate this potential with consideration of a reversibly isomerizing acceptor.

*The relevant expression is (Schachman 1959)

$$\int_{x_m}^{x_p} \Delta\bar{c}x \, dx = \frac{\bar{c}_o x_p^2 x_m^2}{x^2}\left(\ln\frac{x}{x_m}\right)\left(\frac{\Delta\bar{s}}{\bar{s}_{av}}\right).$$

x_m refers to the position of the meniscus, x_p to the point at which Δc reassumes a value of zero, while x refers to the mean boundary position. By observing the pattern at a series of times it is evident that $\Delta\bar{s}/\bar{s}_{av}$ is given by the slope of the plot of the integral quantity against $(\bar{c}_o x_p^2 x_m^2/x^2)\ln(x/x_m)$.

On substituting $n=1$ in eqn (3.2) we obtain

$$u=\frac{v_A+v_C K}{1+K}=\frac{v_A c_A+v_C c_C}{c_A+c_C}, \quad (3.33)$$

the latter equality following from the fact that K is defined on a weight scale ($K=c_C/c_A$). Since the same expression applies to both ascending and descending sides, a sharp boundary moving with weight-average velocity u is therefore observed on each side. By addition of ligand S to the isomerizing acceptor the following situation is established.

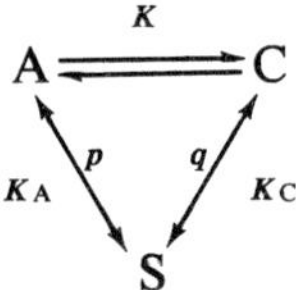

By analogy with eqn (1.23a) we may write

$$\bar{c}_A=M_A m_A(1+K_A m_S)^p, \quad (3.34a)$$

$$\bar{c}_C=M_A m_C(1+K_C m_S)^q, \quad (3.34b)$$

while if v_S is effectively zero (as in velocity sedimentation) then we may assume that $v_A=v_{AS}=\ldots v_{AS_p}$ and $v_C=v_{CS}=\ldots v_{CS_q}$ to a first approximation. The weight-average velocity of acceptor in the presence of ligand is obtained by combining eqns (3.32a, b) with the definition of $\bar{v}$ for this system (eqn (3.35)),

$$\bar{v}_{\text{(plus ligand)}}=\frac{v_A\bar{c}_A+v_C\bar{c}_C}{\bar{c}_A+\bar{c}_C} \quad (3.35a)$$

$$=\frac{v_A+v_C X'}{1+X'}, \quad (3.35b)$$

where

$$X'=\frac{m_C(1+K_C m_S)^q}{m_A(1+K_A m_S)^p}. \quad (3.35c)$$

From eqn (3.35) it is evident that for a system in which K_C and/or $q=0$ an increase in m_S causes the weight-average velocity to approach v_A, while v_C will be approached if K_A and/or $p=0$. Thus, although in one sense A and C are merely labels in the case of isomerizing acceptors, differential experiments comparing the sedimentation rates in the presence and absence of ligand would determine whether it binds preferentially to the isomer with larger or smaller velocity (smaller or larger axial ratio), and hence provide information about the relative shape of the active isomer. In this connection it should be emphasized, however, that such qualitative observations of a change in weight-average velocity on addition of ligand does not establish the validity of adopting the above model involving equivalent and independent sites on two isomeric forms of acceptor. Indeed, a similar $\Delta\bar{s}$ would be observed for an acceptor with equivalent but interacting sites (Koshland and Neet 1968). Thus the information obtained from such experiments is the isomeric or polymeric form of acceptor which binds ligand preferentially, and not the mechanism by which the binding occurs.

3.5. Discussion

From this chapter it is evident that migration experiments provide a reliable qualitative index of self-association in a solute, evidence which, moreover, may be interpreted quantitatively in terms of stoichiometry and association equilibrium constant, provided the system comprises monomer and *single* higher polymer. The value of frontal gel chromatography for such purposes is evident from Fig. 3.6, which emphasizes the low concentrations that can be studied quantitatively by this technique, the lower limit being governed solely by the most sensitive assay for solute. The experimental results for alkaline phosphatase are consistent with a dimerization constant of 10^4 litre/g, and those for sulphatase A with a monomer–tetramer equilibrium and an association constant of $10^{12}\,l^3\,g^{-3}$. In this connection, polymeric species result from bimolecular collisions, and hence the absence of intermediates in systems involving trimers and higher oligomers is at first sight unlikely. Indeed, since we advocate the use of migration experiments only for the quantitative

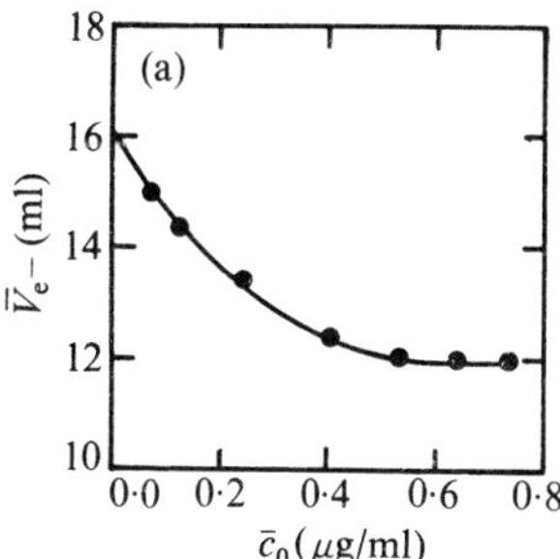

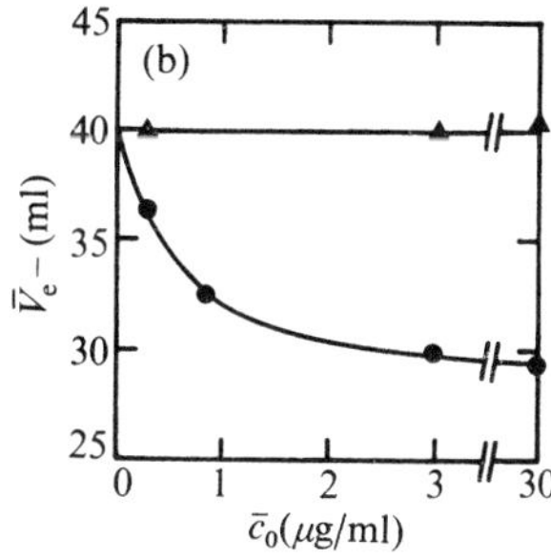

FIG. 3.6. Gel chromatographic evidence of dissociation of enzymes at low protein concentrations: (a) elution of alkaline phosphatase (*E. coli*) from a 1·8 × 11·5 cm column of Sephadex G-100 equilibrated with 1 M tris-HCl, pH 8·0 (adapted from Winzor (1970)); (b) elution of aryl sulphatase A from a 1·5 × 50 cm column of Sephadex G-200 equilibrated with uniunivalent buffers, pH 7·5 (▲), where the enzyme is monomeric, and pH 5·0 (●) (adapted from Nichol and Roy (1965)).

analysis of monomer-single higher polymer systems, one may question the value of such experiments on a solute which contains higher polymers than dimer on these grounds. However, while there are proteins such as insulin (Jeffrey and Coates 1966) which do not conform with the simpler model, there are also several polymerizing systems for which careful experimentation has failed to detect any significant concentrations of intermediate polymers. Perhaps the best-documented example of such a solute is haemerythrin, a monomer–octamer system (Langerman and Klotz 1969). These authors also provide a plausible explanation for the absence of intermediate oligomers by postulating progressively increasing magnitudes for the successive association constants, a free energy difference of as little as 1 kcal between successive polymers sufficing to reduce the concentration of preceding oligomer to an experimentally undetectable level. Such differences in free energies could easily result from minor changes in non-covalent bonding (conformational changes), whereupon the existence of monomer-single higher polymer systems becomes an eminently reasonable postulate.

4

REACTIONS BETWEEN DISSIMILAR MOLECULES

MENTION of the combination of enzymes with substrates and effectors, of antigens with antibodies, and of drugs and hormones with a variety of macromolecular entities suffices to illustrate the importance of studying heterogeneous association reactions. The discussion of the binding of methyl orange to bovine serum albumin, presented in Chapter 2, provides an example of such a study. We commence this chapter by considering further the behaviour of a system undergoing a chemical reaction of the type $A + B \rightleftharpoons AB$ (or C) on being subjected to mass migration in an experiment designed to preserve a plateau region of original solution composition.

4.1. Formation of a single complex

Several cases may be envisaged involving different relative orders of magnitudes of the individual species velocities v_A, v_B, and v_C, and for each of these cases a complete description would require solutions for both ascending and descending sides. One case in which the velocities of the complex and the faster-moving reactant were identical has been considered in detail in Chapter 2. In that development, the convention was adopted that distances were positive in the directions solvent→solution (descending side) and solution→solvent (ascending side), and in these terms the complete solution was illustrated in Fig. 2.4. Let us examine the effect of adopting an alternative convention (Gilbert and Jenkins 1959) which specifies that $v_A \geqq v_B$ and that distances are positive in the direction solvent→solution for *both* ascending and descending sides. If we wish to consider a system in which the faster-moving reactant and complex have identical velocities, for the descending side we examine the case $v_A = v_C > v_B$, the slower-migrating reactant being labelled B: this, of course, is the situation described in the right hand side of Fig. 2.4, and is reproduced in Fig. 4.1(a). According to the new convention, a solution for the conjugate ascending side for the same system is obtained by examining the case $v_A > v_B = v_C$, with the slower reactant labelled A. The form of the boundary distribution is necessarily identical with that shown on the left hand side of Fig. 2.4, which is reproduced in Fig. 4.1(b) with the required relabelling. At first sight, it may appear that nothing has been gained by adopting the new convention, and that, indeed, the approach in Chapter 2 was less confusing. However, consider now a quite different system wherein the velocities of the complex and *slower-moving* reactant are identical. According to the Gilbert–Jenkins convention, the solution for the descending side is obtained by examining the case $v_A > v_B = v_C$, with the slower reactant labelled B, whereupon we find that the solution has already been obtained, being illustrated in Fig. 4.1(b). The conjugate ascending side requires examination of the case $v_A = v_C > v_B$, with the slower reactant labelled A, and is thus given by Fig. 4.1(a).

The value of the Gilbert–Jenkins convention, therefore, is that it obviates unnecessary duplication of certain solutions, and will accordingly be adopted in this chapter, even though it requires that the labels A and B be interchanged for consistency with this convention. With its adoption, the cases that require examination are as follows:

$$\begin{gathered} v_A = v_B = v_C; \\ v_A = v_C > v_B; \quad v_A > v_B = v_C; \\ v_C > v_A = v_B; \quad v_A = v_B > v_C; \\ v_C > v_A > v_B; \quad v_A > v_B > v_C; \\ v_A > v_C > v_B. \end{gathered} \tag{4.1}$$

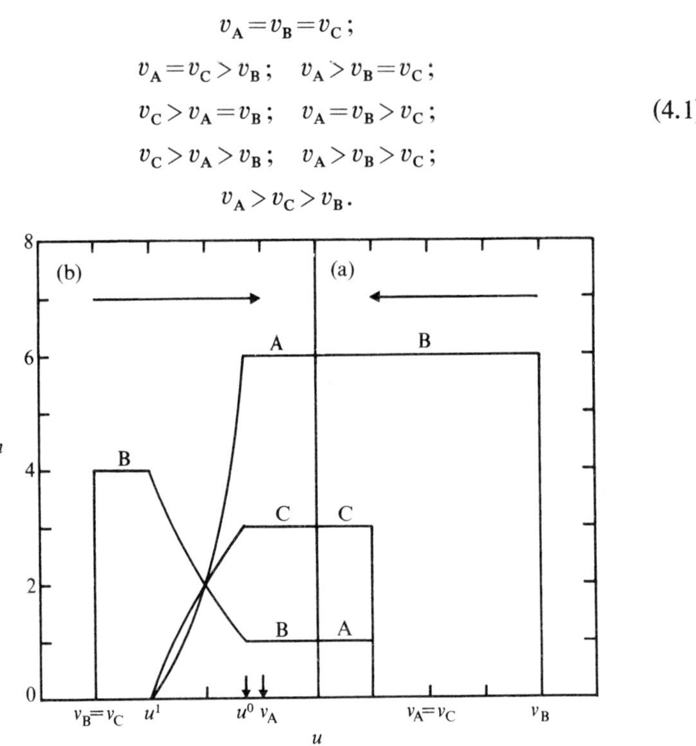

FIG. 4.1. A reproduction of Fig. 2.4 with the relabelling required by the convention that $v_A \geqq v_B$ with distances positive in the direction solvent→solution for both sides. Pattern (a) refers *either* to a descending side for the case $v_A = v_C > v_B$, with the slower reactant labelled B, *or* to an ascending side for the case $v_A = v_C > v_B$, with slower reactant labelled A. Pattern (b) refers *either* to a descending side for the case $v_A > v_B = v_C$, with slower reactant labelled B, *or* to an ascending side for the case $v_A > v_B = v_C$, with slower reactant labelled A.

For each of these cases the appropriate constituent concentrations and velocities are defined by eqns (2.9a, b, c, d), where the individual species concentrations m_A, m_B, and m_C are expressed on a molar basis and the individual species velocities are taken as constants. If, in addition, the assumptions specified in Chapters 2 and 3 (viz. rapid re-equilibration, a rectangular cell, neglect of diffusional flows, etc.) continue to apply, we may use eqns (2.10) and (2.11a, b) jointly to deduce the forms of migration patterns. It will also be assumed, for the present, that u_A and u_B as defined by eqns (2.11a) and (2.11b), respectively, are identical, this having been done also in Chapter 2. Equations (2.9a, b, c, d), (2.10), and (2.11a, b) may be combined or rewritten to eliminate constituent quantities to give

$$(u-v_A)\, dm_A = -(u-v_C)\, dm_C \tag{4.2a}$$

$$(u-v_B)\, dm_B = -(u-v_C)\, dm_C \tag{4.2b}$$

$$K = \frac{m_C}{m_A m_B}, \tag{4.3a}$$

$$Km_A\, dm_B + Km_B\, dm_A = dm_C. \tag{4.3b}$$

Of course, these equations are identical with eqns (2.37a, b) and (2.38); they have been rewritten here for easy reference. For the case $v_A = v_B = v_C$ (first line of condition (4.1), it follows from eqn (4.2a, b) that when $u \neq v_A$, *either* $dm_A = dm_B = dm_C = 0$ *or* dm_A and dm_B are of the same sign but opposite to that of dm_C. The latter possibility is inconsistent with eqn (4.3b), and thus $dm_A = dm_B = dm_C$ except when $u = v_A = v_B = v_C$. There is therefore only a single, sharp boundary at $v_A t$, separating the solvent plateau and the original solution plateau. Ascending and descending sides are enantiographic, the behaviour exactly paralleling that of a single solute described in eqn (2.5). The cases specified in the second line of condition (4.1) have been considered already in Chapter 2, the results being summarized in Fig. 4.1. Cases in the third line of condition (4.1) form an interesting set in that they may be correlated in special circumstances with a dimerizing system, and thus provide a link with Chapter 3. We shall develop the solution of the non-enantiographic migration patterns for these cases and thereby stress the general utility of eqns (4.2) and (4.3) and the Gilbert–Jenkins convention.

(a) *The migration of a system involving two reactants with equal velocities: an analogy with a dimerizing system*

$v_C > v_A = v_B$

The case $v_C > v_A = v_B$ corresponds to a situation in a descending side where two reactants with identical velocities move slower than the complex, or to an ascending side where, with a quite different system, the absolute magnitude of the velocity of the complex is less than that of the reactants. It will be required later to obtain solutions for the conjugate sides applicable to both of these systems. In the discussion of $v_C > v_A = v_B$ reference is made to Fig. 4.2(a).

(i) When $u < v_A$, v_B, and v_C, the only solution of eqns (4.2a, b) and (4.3b) is $dm_A = dm_B = dm_C = 0$, which necessarily describes the solvent plateau.

(ii) When $u = v_A = v_B$, eqns (4.2a, b) show that $dm_C = 0$ since $(u - v_C) \neq 0$. At this position ($v_A t$), m_C continues to equal zero, but eqns (4.2a, b) place no

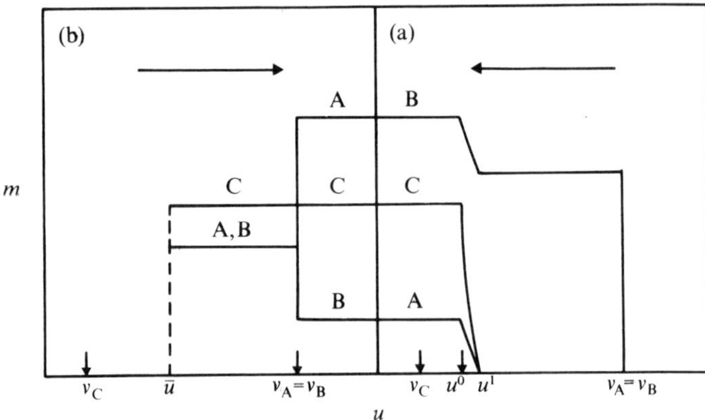

FIG. 4.2. Representative patterns for systems involving reactants with equal velocities ($v_A = v_B$), the broken line indicating a hypersharp boundary. (a) *Either* a descending side for the case $v_C > v_A = v_B$, *or* an ascending side for the same case: the numerical example specifies that $m_{o,B} > m_{o,A}$. (b) *Either* a descending side for the case $v_A = v_B > v_C$, *or* an ascending side for the same case: with the appropriate relabelling it now follows that $m_{o,A} > m_{o,B}$. (Adapted from Gilbert and Jenkins (1959).)

restriction on dm_A and dm_B: eqn (4.3a) does prohibit the possibility that $dm_A \neq 0$ and $dm_B \neq 0$. The following possibilities remain: (a) $dm_A = 0$ and $dm_B \neq 0$ (i.e. a sharp boundary of B arises at $v_B t$ to a level m'_B); (b) $dm_A \neq 0$ and $dm_B = 0$ (i.e. a sharp boundary of A arises at $v_A t$ to a level m'_A); and (c) $dm_A = 0$ and $dm_B = 0$ (i.e. no boundary arises at $v_A t = v_B t$). It will be shown later that any of these three possibilities may pertain, depending on the composition of the original solution subjected to mass migration: but it suffices in developing this point to select one possibility, and we choose (a), as illustrated in Fig. 4.2(a).

(iii) When $u > v_A = v_B$, we examine the combination of eqns (4.2a) and (4.3b), viz.

$$\{(u - v_A) + (u - v_C)Km_B\}\, dm_A + (u - v_C)Km_A\, dm_B = 0. \tag{4.4}$$

One solution is that $m_A = 0$ and $dm_A = 0$, which describes a plateau of B alone of concentration m'_{B}, the value of which remains to be specified. This plateau is terminated at a point of singularity $u't$ at which m_{A} commences to increase from zero. An expression for u' follows by placing $m_{\mathrm{A}}=0$ in eqn (4.4) and rearranging the solution (other than $\mathrm{d}m_{\mathrm{A}}=0$) of the resultant expression.

$$u'=\frac{v_{\mathrm{A}}+v_{\mathrm{C}}Km_{\mathrm{B}}}{1+Km_{\mathrm{B}}}. \tag{4.5}$$

When $u>u'$ and $\mathrm{d}m_{\mathrm{A}}>0$ (i.e. m_{A} is increasing from zero in the presence of B), eqns (4.3a, b) require that m_{C} must also increase from zero, i.e. $\mathrm{d}m_{\mathrm{C}}>0$ and, therefore, $\mathrm{d}m_{\mathrm{A}}/\mathrm{d}m_{\mathrm{C}}>0$. In this region it follows from eqns (4.2a, b) that

$$\frac{\mathrm{d}m_{\mathrm{A}}}{\mathrm{d}m_{\mathrm{C}}}=\frac{\mathrm{d}m_{\mathrm{B}}}{\mathrm{d}m_{\mathrm{C}}}=-\frac{(u-v_{\mathrm{C}})}{(u-v_{\mathrm{A}})}>0, \tag{4.6}$$

which shows that when $\mathrm{d}m_{\mathrm{A}}$ and $\mathrm{d}m_{\mathrm{C}}>0$ then also $\mathrm{d}m_{\mathrm{B}}>0$, and also that all values of u are greater than v_{A} and less than v_{C} in the region where these inequalities apply. It is also clear that $\mathrm{d}m_{\mathrm{A}}/\mathrm{d}m_{\mathrm{B}}=1$, a relationship which may be integrated between the plateau in the region $u<u'$ ($m_{\mathrm{B}}=m'_{\mathrm{B}}$, $m_{\mathrm{A}}=m_{\mathrm{C}}=0$) and the original solution plateau ($m_{\mathrm{B}}=m_{\mathrm{o,B}}$, $m_{\mathrm{A}}=m_{\mathrm{o,A}}$, $m_{\mathrm{C}}=m_{\mathrm{o,C}}$) to give

$$m'_{\mathrm{B}}=m_{\mathrm{o,B}}-m_{\mathrm{o,A}}\equiv\Delta. \tag{4.7}$$

More generally, the integration of $\mathrm{d}m_{\mathrm{A}}/\mathrm{d}m_{\mathrm{B}}=1$ in the boundary region relates the corresponding values of m_{A} and m_{B} by

$$m_{\mathrm{B}}=\Delta+m_{\mathrm{A}}. \tag{4.8}$$

Substitution of eqn (4.8) and $\mathrm{d}m_{\mathrm{A}}=\mathrm{d}m_{\mathrm{B}}\neq 0$ into eqn (4.4) gives

$$m_{\mathrm{A}}=\frac{-(u-v_{\mathrm{A}})-(u-v_{\mathrm{C}})K\Delta}{2K(u-v_{\mathrm{C}})} \tag{4.9}$$

or, with introduction of the original solution concentrations of A and B,

$$u^{\mathrm{o}}=\frac{v_{\mathrm{A}}+Kv_{\mathrm{C}}(m_{\mathrm{o,B}}+m_{\mathrm{o,A}})}{1+K(m_{\mathrm{o,B}}+m_{\mathrm{o,A}})}. \tag{4.10}$$

The solution is complete, and any numerical example may be constructed for a mixture of defined composition (which fixes $m_{\mathrm{o,A}}$, $m_{\mathrm{o,B}}$, $m_{\mathrm{o,C}}$ and hence K and Δ) and for selected values of $v_{\mathrm{A}}=v_{\mathrm{B}}$ and v_{C}. As is seen in Fig. 4.2(a), m_{B} increases at $ut=v_{\mathrm{B}}t$ to a value m'_{B} given by eqn (4.7): a plateau of B alone at this concentration extends to $u't$ (given by eqn (4.5)), where a spread boundary in A, B, and C commences; the exact form of the spread boundary is described by eqns (4.9), (4.8), and (4.3a), it being terminated by the original solution plateau at $u^{\mathrm{o}}t$, given by eqn (4.10).

From eqn (4.7) it is clear that the above description only applies when $m_{o,B} > m_{o,A}$. In the event that $m_{o,A} > m_{o,B}$, a sharp boundary in A (rather than B) arises at $v_A t = v_B t$ (possibility (b) above) to a level $m'_A = m_{o,A} - m_{o,B} \equiv \Delta_A$, whereupon a complete solution may be obtained from eqns (4.5), (4.9), and (4.10) with the substitution of m'_A for m'_B and of Δ_A for Δ.

For the special case that $m_{o,A} = m_{o,B}$ it is clear that $\Delta = \Delta_A = 0$, i.e. $dm_A = dm_B = 0$ at $v_A t = v_B t$ (possibility (c) above). Thus only a single spread boundary appears between the limit $u't = v_A t$ (from eqn (4.5)) and the position $u^{o}t$ (given by eqn (4.10)). Also, when $\Delta = 0$, eqns (4.8) and (4.9) give

$$m_A = m_B = -\frac{(u - v_A)}{2K(u - v_C)}. \tag{4.11}$$

Since $v_A = v_B$ and $m_{o,A} = m_{o,B}$, this particular $A + B \rightleftharpoons C$ system (with $M_A = M_B$) is equivalent to a dimerization equilibrium. Placing $n = 2$ in eqn (3.2) and rearranging, we find

$$c_A \text{ (g/l)} = -\frac{(u - v_A)}{2(u - v_C)K \text{ (l/g)}} \tag{4.12a}$$

or

$$m_A = -\frac{(u - v_A)}{(u - v_C)K_2 \text{ (l/mol)}}. \tag{4.12b}$$

It is seen that eqns (4.12b) and (4.11) are consistent on noting that m_A in the former is equivalent to $(m_A + m_B)$ in eqn (4.11): each describes explicitly a single, spread reaction boundary with starting limit $v_A t$.

$v_A = v_B > v_C$

This case corresponds to a situation in a descending side where the reactants with identical velocities move faster than the complex, or to an ascending side where the absolute magnitude of the velocity of the complex is greater than that of the reactants. The solution will thus provide a description of the conjugate sides for the cases considered above ($v_C > v_A = v_B$). Reference is therefore made to Fig. 4.2(b).

(i) When $u > v_B$, $(u - v_A)$, $(u - v_B)$, and $(u - v_C) \neq 0$, whereupon eqns (4.2a, b) are satisfied *either* by $dm_A = dm_B = dm_C = 0$ (plateau of original composition) *or* by $dm_A = dm_B$, $dm_A/dm_C < 0$, and $dm_B/dm_C < 0$ (which is incompatible with eqn (4.3b).

(ii) When $u = v_A = v_B$, eqns (4.2a, b) require that $dm_C = 0$, since $(u - v_C) \neq 0$. Thus at $v_A t = v_B t$, $m_C = m_{o,C}$, which is an entirely different situation from that encountered at the same position on the conjugate side. Equation (4.3b) now becomes $m_A\,dm_B + m_B\,dm_A = 0$, and two possibilities arise: (a) $dm_A = dm_B = 0$, and (b) $dm_A/dm_B = -m_A/m_B < 0$. In either case, $d(m_A m_B) = 0$ at this position, and integration to the original solution plateau gives

$$m'_A m'_B = m_{o,A} m_{o,B} = \Pi \neq 0. \tag{4.13}$$

This is a necessary condition inter-relating the plateaux; but it is not a sufficient condition for the explicit description of the behaviour at the position $v_A t$. Considerations of symmetry and physical reasoning suggest (Gilbert and Jenkins 1959, Nichol and Ogston 1965*a*) that if $dm_A/dm_B < 0$ at $v_A t$, then, when $m_{o,A} > m_{o,B}$, m_A decreases across the sharp boundary at this position, while m_B increases (thereby ensuring that $dm_A/dm_B < 0$) until $m'_A = m'_B = \Pi^{1/2}$ (Fig. 4.2(b)). Alternatively, when $m_{o,B} > m_{o,A}$, the situation would be reversed. In the event that $m_{o,A} = m_{o,B}$ there would certainly be no change in m_A and m_B at $v_A t$ (possibility (a) above, viz. $dm_A = dm_B = 0$).

(iii) When $u = v_C$ and $(u - v_A) = (u - v_B) \neq 0$, eqns (4.2a, b) and (4.3b) considered together insist that $dm_A = dm_B = dm_C = 0$, which can only reasonably describe the solvent plateau. Thus solute concentrations must change from zero to m'_A, m'_B, $m_{o,C}$ somewhere in the region $v_B > u > v_C$.

(iv) For at least part of the region $v_B > u > v_C$ we know that $m'_A = m'_B$ and $dm_A = dm_B$, whereupon eqn (4.4) becomes

$$\{(u - v_A) + (u - v_C)2Km_B\}\,dm_B = 0. \tag{4.14}$$

One solution, $dm_B = 0$, describes the plateau $m'_A = m'_B$, $m_C = m_{o,C}$. The remaining solution is $(u - v_A) + (u - v_C)2Km_B = 0$, which gives on differentiation with respect to u

$$\frac{dm_B}{du} = \frac{dm_A}{du} = \frac{(v_C - v_B)}{2K(u - v_C)^2} < 0. \tag{4.16}$$

It follows from eqn (4.3b) that $dm_C/du < 0$ and hence that $d\bar{m}_A/du$ and $d\bar{m}_B/du$ are negative. At the same time, with the adopted convention, $d\bar{m}_A/dx$ and $d\bar{m}_B/dx$ are positive, and consequently the boundary separating the solvent plateau and that wherein $m'_A = m'_B$, $m_C = m_{o,C}$ is hypersharp (at least according to the rule adopted by Nichol and Ogston 1965*a*): this hypersharp boundary is represented in Fig. 4.2(b) by the broken line situated at $\bar{u}t$. The value of $\bar{u}$ follows from integration of either (2.11a) or (2.11b) between the plateaux which the hypersharp boundary separates. With the assumption that u_A and u_B are identical, integration of eqns (2.11a, b) gives

$$\bar{u} = \bar{v}_A = \bar{v}_B = \frac{v_A m'_A + v_C m_{o,C}}{m'_A + m_{o,C}} = \frac{v_B m'_B + v_C m_{o,C}}{m'_B + m_{o,C}} \tag{4.17a}$$

$$=\frac{v_A+v_C K\Pi^{1/2}}{1+K\Pi^{1/2}}. \tag{4.17b}$$

It may also be noted that $\bar{u}$ is a weight-average velocity of all species in the plateau wherein $m'_A = m'_B$, $m_C = m_{o,C}$.

In the special case that $m_{o,A} = m_{o,B}$ and $M_A = M_B$, the required condition expressed in eqn (4.13) is met without the necessity of a change in the concentrations of A and B at $v_A t$: thus only a single, hypersharp boundary moving with the weight-average velocity corresponding to the original plateau concentrations would be observed: this is entirely consistent with the behaviour derived in Chapter 3 for a dimerizing system on the side where $v_C < v_A$ (distances positive in the direction solvent→solution). Even with this special case, then, conjugate patterns are non-enantiographic, one being hypersharp and the other spread. In this connection, it is noted that differentiation of eqns (4.8) and (4.9) referring to Fig. 4.2(a) yields eqn (4.16), with the exception that, since $v_C > v_B$ on this side, dm_B/du and dm_A/du (and hence $d\bar{m}_B/du$ and $d\bar{m}_A/du$) are positive, as are $d\bar{m}_A/dx$ and $d\bar{m}_B/dx$.

Practical application of Fig. 4.2. The question is raised whether a migration pattern such as that illustrated in Fig. 4.2 may be used to determine the equilibrium constant K governing an $A+B \rightleftharpoons C$ system for which $v_A = v_B$. The composition of the original solution may be determined in terms of the total concentrations of A and B, $\bar{m}_A$ and $\bar{m}_B$, respectively, while the equilibrium constant may be written using eqns (2.9a, b) as

$$K=\frac{m_{o,C}}{m_{o,A}m_{o,B}}=\frac{(\bar{m}_{o,B}-m_{o,B})}{m_{o,B}(\bar{m}_{o,A}-\bar{m}_{o,B}+m_{o,B})}=\frac{(\bar{m}_{o,A}-m_{o,A})}{m_{o,A}(\bar{m}_{o,B}-\bar{m}_{o,A}+m_{o,A})}. \tag{4.18}$$

Hence the value of K may be determined if the equilibrium concentration of unbound A or B *in the original solution plateau* may be determined. It is clear from Fig. 4.2(a) and eqn (4.7) that the concentration of B in the region where it exists alone is not the appropriate equilibrium concentration of B in the original solution plateau. The situation may be contrasted with that shown in Fig. 4.1(a) (pertaining to $v_A = v_C > v_B$), where $m_{o,B}$ pertinent to eqn (4.18) may be measured directly.

An alternative possibility for determining K for a system where $v_A = v_B$ follows from inspection of Fig. 4.2(b). The value of $\bar{u}$ may be determined with some precision, since the boundary is hypersharp. Rearrangement of eqn (4.17b) gives

$$\left\{\frac{v_A-\bar{u}}{\bar{u}-v_C}\right\}^2 = K^2\Pi \equiv Z, \tag{4.19}$$

from which Z may be evaluated if v_A and v_C are known. Now, using eqn (4.13), we find

$$K = \frac{m_{o,C}}{m'_A m'_B} = \frac{m_{o,C}}{\Pi}. \tag{4.20}$$

Combination of eqns (4.19) and (4.20) shows that $m_{o,C} = Z/K$, whereupon

$$K = \frac{Z/K}{(\bar{m}_{o,A} - Z/K)(\bar{m}_{o,B} - Z/K)} \tag{4.21a}$$

or

$$\frac{1}{K^2}(Z - Z^2) + \frac{1}{K}(Z\bar{m}_{o,A} + Z\bar{m}_{o,B}) - \bar{m}_{o,A}\bar{m}_{o,B} = 0. \tag{4.21b}$$

Only one solution of this quadratic expression in K is physically acceptable, the other predicting the unreal situation that $m_{o,C} > \bar{m}_{o,A}$ or $\bar{m}_{o,B}$. Nichol and Winzor (1965) applied this method to frontal analysis results obtained with the trypsin-soybean trypsin inhibitor system on Sephadex G-100. The elution volumes of trypsin and the inhibitor studied separately were found to be virtually identical in the pH range 4·25–5·50 (thereby establishing $v_A = v_B$). Values of $\bar{u}$ varied with pH, that obtained at pH 5·5 being taken as a measure of v_C since Lebowitz and Laskowski (1962) had shown that the equilibrium was shifted completely toward the 1 : 1 complex at this pH. Unfortunately, Nichol and Winzor (1965) incorrectly used reciprocals of elution volumes in their calculations of Z: Gilbert (1966*a*) corrected this error, and thereby showed that K values obtained from the results of Nichol and Winzor (1965) and Lebowitz and Laskowski (1962) were in excellent agreement.

In summary, a solution has been obtained for the migration of two reactants with identical velocities. It bears relation to the behaviour of a dimerizing system, and the properties of the hypersharp boundary and the plateaux which it separates may be used to determine K. However, the determination of the equilibrium constant for such a system requires independent evaluation of v_A and v_C, parameters which were not needed in the treatment of results for a system in which $v_C = v_A > v_B$ (Fig. 4.1(a)).

(b) *Cases involving no equalities between velocities of the individual species*

$v_C > v_A > v_B$

This case corresponds to a descending side where both reactants have velocities less than that of the complex, or to an ascending side where the complex has a smaller absolute migration rate than either reactant. Reference is made to Fig. 4.3(a), (b). Provided it is assumed that u_A and u_B of eqns (2.11a) and (2.11b), respectively, are identical, eqns (4.2a, b) apply together with eqns (4.3a, b): they may be used, as before, to show that the solvent plateau extends when $u < v_B$ and that the original solution plateau extends when $u \geqq v_C$. When $u = v_B$ it is possible that $dm_B \neq 0$, $dm_A = 0$, $dm_C = 0$ (a sharp boundary in B arises as illustrated in Fig. 4.3(a)), or that $dm_B = dm_A = dm_C$ (the solvent plateau continues). Let us consider the former possibility, and note that in the region $v_C > v_A > u > v_B$, $(u - v_A) < 0$, $(u - v_B) > 0$, and $(u - v_C) < 0$. Unless $dm_A = dm_B = dm_C = 0$, it would be required by eqns (4.2a, b) that $dm_A/dm_B < 0$ and $dm_A/dm_C < 0$, which is impossible. Thus a plateau of pure B continues until $u > v_A$: when $u = v_A$, dm_C (and hence dm_A) is zero. Similarly, it may be shown

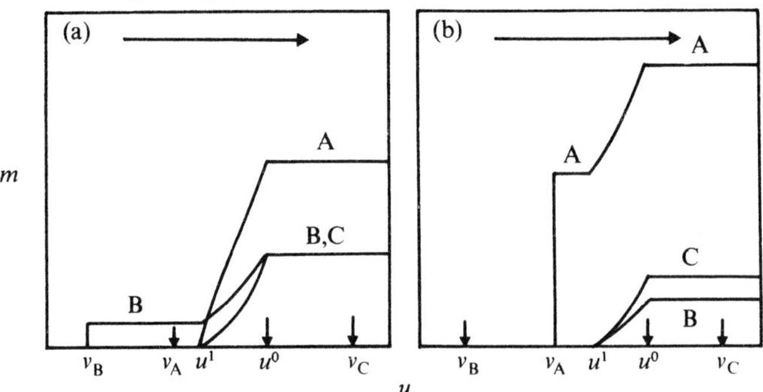

FIG. 4.3. Representative migration patterns for the case $v_C > v_A > v_B$. With distances positive in the direction solvent→solution both patterns correspond *either* to a descending side where both reactants have velocities less than that of the complex, *or* to an ascending side where the complex has a smaller absolute migration rate than that of either reactant: (a) $\Phi > 0$; (b) $\Phi < 0$ (see text). (Adapted from Gilbert and Jenkins (1959).)

that if at $v_B t$ the equality $dm_B = dm_A = dm_C = 0$ holds, a sharp boundary in A may arise at $v_A t$, as illustrated in Fig. 4.2(b). In either case it is only in the region $v_C > u > v_A > v_B$ that dm_A, dm_B, and dm_C may all be positive, a situation which describes the reaction boundary separating the original solution plateau from that containing either A or B alone. The approach, thus far, has indicated the general forms of the boundary distributions that may occur; but has left several questions unanswered. Gilbert and Jenkins (1959) provided a complete solution, which will now be formulated on the basis of eqns (4.2a, b) and (4.3a, b).

Examination of the reaction boundary in the region $v_C > u > v_A$ is facilitated by introducing the following relationships:

$$K = \frac{1}{k}, \tag{4.22a}$$

$$\frac{dm_A}{dm_B}=\frac{(u-v_B)}{(u-v_A)}=\mu, \tag{4.22b}$$

$$\frac{dm_C}{dm_B}=\frac{-(u-v_B)}{(u-v_C)}=\omega, \tag{4.22c}$$

$$\lambda=\frac{2v_C-(v_A+v_B)}{(v_A-v_B)}, \tag{4.22d}$$

$$(\lambda+1)=\frac{2(v_C-v_B)}{(v_A-v_B)}, \tag{4.22e}$$

$$(\lambda-1)=\frac{2(v_C-v_A)}{(v_A-v_B)}, \tag{4.22f}$$

$$\mu(\lambda-1)-(\lambda+1)=\frac{-2(u-v_C)}{(u-v_A)}. \tag{4.22g}$$

In these terms, eqn (4.3b) becomes $m_A=k\omega-m_B\,\mu$, which may be differentiated with respect to m_B to give

$$2\mu=k\frac{d\omega}{dm_B}-m_B\frac{d\mu}{dm_B}, \tag{4.23a}$$

or

$$2\mu\frac{dm_B}{d\mu}+m_B=k\frac{d\omega}{d\mu}, \tag{4.23b}$$

or

$$2\sqrt{\mu}\,\frac{dm_B\sqrt{\mu}}{d\mu}=k\frac{d\omega}{d\mu}. \tag{4.23c}$$

It also follows from differentiating eqns (4.22b, c) that

$$\frac{d\omega}{d\mu}=-\left(\frac{u-v_A}{u-v_C}\right)^2\cdot\left(\frac{v_C-v_B}{v_A-v_B}\right) \tag{4.24a}$$

$$=-\frac{2(\lambda+1)}{\{(\lambda-1)\mu-(\lambda+1)\}^2}, \tag{4.24b}$$

the latter equality following directly from eqns (4.22e, g). Combination of eqns (4.23c) and (4.24b) gives eqn 14 of Gilbert and Jenkins (1959), which they wrote in integrated form as

$$\frac{m_B}{k}=-\frac{(\lambda+1)}{\sqrt{\mu}}\int\frac{d\mu}{\sqrt{\mu}\{(\lambda-1)\mu-(\lambda+1)\}^2} \tag{4.25}$$

At this stage Gilbert and Jenkins (1959) introduced an elegant substitution to permit evaluation of the integral.

$$u=\frac{(\lambda+1)}{(\lambda-1)}\coth^2\tfrac{1}{2}\theta \tag{4.26a}$$

where θ at a particular value of u is given by

$$\tanh\tfrac{1}{2}\theta=\frac{(e^{1/2\,\theta}-e^{-1/2\,\theta})}{(e^{1/2\,\theta}+e^{-1/2\,\theta})} \tag{4.26b}$$

$$=\left\{\frac{(\lambda+1)\,(u-v_A)}{(\lambda-1)\,(u-v_B)}\right\}^{1/2} \tag{4.26c}$$

Substitution of eqn (4.26a) into eqn (4.25) yields, after extensive rearrangement,

$$\frac{m_B}{k}=\frac{\tanh\frac{1}{2}\theta}{(\lambda+1)}\int\frac{\operatorname{csch}^2\frac{1}{2}\theta\,d\theta}{(\coth^2\frac{1}{2}\theta-1)^2} \tag{4.27a}$$

$$=\frac{\tanh\frac{1}{2}\theta}{(\lambda+1)}\int\sinh^2\tfrac{1}{2}\theta\,d\theta \tag{4.27b}$$

$$=\frac{\tanh\frac{1}{2}\theta}{2(\lambda+1)}\int(\cosh\theta-1)\,d\theta \tag{4.27c}$$

$$=\frac{\sinh^2\frac{1}{2}\theta}{(\lambda+1)}\left(1+\frac{\Phi-\theta}{\sinh\theta}\right), \tag{4.27d}$$

where Φ is the constant of integration. An entirely similar approach may be used to obtain an expression relating the equilibrium concentration of A in the reaction boundary region to θ, the function of velocities (eqn (4.26c)):

$$\frac{m_A}{k}=\frac{1}{(\lambda-1)}\left(1-\frac{\Phi-\theta}{\sinh\theta}\right)\cosh^2\tfrac{1}{2}\theta. \tag{4.28}$$

Combinations of eqns (4.3a), (4.22a), (4.27d), and (4.28) yields the following expression for m_C/k:

$$\frac{m_C}{k}=\frac{1}{(\lambda^2-1)}\left(1-\frac{(\Phi-\theta)^2}{\sinh^2\theta}\right)\frac{\sinh^2\theta}{4}. \tag{4.29}$$

The construction of a numerical example commences by assigning values to the equilibrium concentrations $m_{o,A}$, $m_{o,B}$, and $m_{o,C}$, pertinent to the original solution plateau (this being equivalent to selecting K, $\bar{m}_A$, and $\bar{m}_B$, as is evident from eqn (4.18)). Equations (4.27d) and (4.28) apply to the point of singularity $u^o t$, where the original solution plateau is terminated by the reaction boundary, and the equations may be solved simultaneously (for known values of $m_{o,A}$, $m_{o,B}$, k, v_A, v_B, v_C, and hence λ) to obtain values of Φ

and θ°. The latter quantity may then be used in eqn (4.26c) to obtain the value of u°. The reaction boundary, whose form may be computed from eqns (4.26c), (4.27d), (4.28), and (4.29), is terminated on the other side at a point of singularity $u't$, where, as we have seen, *either* $m_A=0, m_C=0, m_B\neq0$ *or* $m_A\neq0, m_B=0, m_C=0$. Gilbert and Jenkins (1959) noted that the former situation applies when Φ, the constant of integration, is positive, whereupon, with $m_A=0$ in eqn (4.28), we find that

$$\sinh\theta'+\theta'=\Phi \qquad (\Phi>0) \tag{4.30a}$$

and the value of θ' thus calculated may be used in eqn (4.26c) to give u'. Substitution of eqn (4.30a) into eqn (4.27d) yields

$$\frac{m'_B}{k}=\frac{2}{(\lambda+1)}\sinh^2\tfrac{1}{2}\theta', \tag{4.30b}$$

which gives the concentration of B in the plateau containing it alone: this plateau is terminated by the solvent plateau at $v_B t$ (Fig. 4.3(a)). Alternatively, if $\Phi<0$, then $m_A\neq0, m_B=0, m_C=0$, and eqn (4.27d) gives (with $m_B=0$ at $u't$)

$$\sinh\theta'-\theta'=|\Phi| \qquad (\Phi<0) \tag{4.31a}$$

while eqns (4.28) and (4.31a) yield

$$\frac{m'_A}{k}=\frac{2}{(\lambda-1)}\cosh^2\tfrac{1}{2}\theta', \tag{4.31b}$$

which completes the solution illustrated in Fig. 4.3(b). It is also possible to differentiate eqns (4.27d), (4.28), and (4.29) to provide a solution in differential form akin to a schlieren pattern.

Practical application of Fig. 4.3. The patterns illustrated in Fig. 4.3 may well apply to results obtained in sedimentation velocity, where heterogeneous association of two reactants characterized by different sedimentation coefficients leads to the formation of a complex with an even higher sedimentation coefficient. In these cases, the trailing boundary will represent pure reactant, either A or B, and a distinction between the possibilities may be made by comparing the velocity of the slower-moving boundary (either v_A or v_B) with those of the individual reactants: confirmation by sampling and subsequent analysis is desirable. This is particularly important because Φ may be greater or less than zero for a given system, depending on the values of $Km_{o,A}$ and $Km_{o,B}$. Thus for a system characterized by a particular value of K, a variation of the initial mixing proportions over a large range may result in the replacement of a boundary characteristic of one reactant by one related to the other reactant. Gilbert and Jenkins (1963) illustrate this point further in their prediction of model sedimentation velocity patterns for the system pepsin + bovine serum albumin $\rightleftharpoons$ complex. The behaviour is not unexpected on physical grounds. A reactant present in smaller amount would be partially removed by complex formation, and the removal of complex in the migration process then provides opportunity for the less concentrated reactant to be depleted further by more complex formation. The continuation of the process must eventually lead to a region containing only the reactant present in excess, moving behind the reaction boundary. This reaction boundary may appear bimodal in a schlieren pattern, as seems to be the case in the sedimentation velocity of a system comprising univalent antigen (bovine serum mercaptalbumin with one benzenearsonic acid residue covalently bonded per molecule) and a bivalent antibody (Pepe and Singer 1956, 1959; Singer, Pepe, and Ilten 1959). The positions of termination u' and u° (which would be affected by diffusion in practice) lie between v_A and v_C.

Perhaps the most important feature in Fig. 4.3, which also follows from eqns (4.30b) and (4.31b), is that the concentration of reactant (albeit A or B) in the plateau of pure solute does not equal the corresponding equilibrium concentration of that reactant in the original solution plateau. Therefore, regardless of the accuracy with which this plateau concentration of reactant is determined (for example, by absorption optics), the value cannot be used directly with $\bar{m}_A$ and $\bar{m}_B$ in eqn (4.18) to compute K. The method of sedimentation dialysis, as described by Blake and Peacocke (1968), must therefore be used with caution in interpreting systems for which $v_C > v_A > v_B$. It is possible to use the value of m'_B or m'_A obtained in these circumstances to compute the value of $m_{o,B}$ or $m_{o,A}$ appropriate to eqn (4.18), but the integrated form of the continuity equation required (Longsworth 1959*a*, Nichol and Winzor 1964) involves velocity terms that may only be assessed approximately (see Gilbert and Kellett (1971) and Section 5b of Chapter 5). It is evident from Fig. 4.1(a) that the difference between (say) m'_B and $m_{o,B}$ decreases as v_C approaches v_A, so that when $v_C = v_A$ the value of m'_B $(= m_{o,B})$ may be used directly in eqn (4.18) without the necessity of estimating velocity terms. In this sense, the method of frontal analysis in gel chromatography, as described in Chapter 2, is to be preferred to sedimentation dialysis for the determination of association constants for equilibria of the type A + B $\rightleftharpoons$ C.

$v_A > v_B > v_C$

This case corresponds to the situation on a descending side for a system in which both reactants have velocities greater than the complex, or to an ascending side, where the complex has a greater absolute migration rate than that of either reactant. The solutions for this case, therefore, pertain to the conjugate sides of those described in the previous section. A general description of the possible migration patterns has been given by Nichol and Ogston (1965*a*), who utilized eqns (4.2a, b) and (4.3a, b). We shall not duplicate their detailed reasoning here, since the approach has already been amply illustrated with other cases. Moreover, the present case is not likely to be encountered frequently in practice, and it therefore suffices to refer briefly to two possible solutions shown in Fig. 4.4(a)(b). That basically two solutions are possible is shown by combining eqns (4.2b) and (4.3b) to give

$$(u-v_B)\,dm_B+K(u-v_C)\,dm_A m_B=0. \tag{4.32}$$

In the region $v_A>v_B>u>v_C$, where $(u-v_A)<0$, $(u-v_B)<0$, and $(u-v_C)>0$, eqn (4.32) shows that $dm_A m_B>0$, since dm_B is necessarily positive. One possibility is that $m_A=0$, $dm_A=0$, $dm_B>0$, whereupon eqn (4.32) has only one solution, viz. $u=v_B$ (i.e. a sharp boundary in B arises at $v_B t$ as illustrated in Fig. 4.4(a)). The other possibility is that $dm_A \neq 0$, $dm_B \neq 0$, which, as shown by Nichol and Ogston (1965*a*), is in accord with a hypersharp boundary, illustrated in Fig. 4.4(b) (i.e. for this boundary $d\bar{m}_A/dx$ is positive and $d\bar{m}_A/du$ is negative). On returning to Fig. 4.4(a), we find the sharp boundary of B adjacent to a spread reaction boundary. A *complete* solution for this possibility was obtained previously by Gilbert and Jenkins (1959), and included expressions for the following parameters (equation numbers in parentheses being those of Gilbert and Jenkins): m'_B (eqn (24)); u' (eqns (22) and (23)); u^0 (eqns (21a), (21b), and (22)); m_A, m_B, and m_C in the reaction boundary region as a function of u (eqns (21a, b, c) and (22)). The method of using these equations is similar to that described in the treatment of our eqns (4.26c), (4.27d), (4.28), (4.29), and (4.30a, b).

The second possibility, illustrated in Fig. 4.4(b) and deduced by both groups of workers, specifies a hypersharp boundary adjacent to a spread reaction boundary: again, Gilbert and Jenkins (1959) have provided an analytical solution describing all features of the pattern. A third possibility (Fig. 6(c) of Gilbert and Jenkins) suggested the existence of two adjacent hypersharp boundaries. Nichol and Ogston (1965*a*) agreed with this possibility originally; but later pointed out that there was doubt as to its validity (Nichol and Ogston 1967). The difficulty is that, while it is possible to show that two spread boundaries formed by the same set of constituents cannot be separated by a common plateau region, it has not yet been possible to show whether or not a similar limitation applies to the adjacency of hypersharp boundaries.

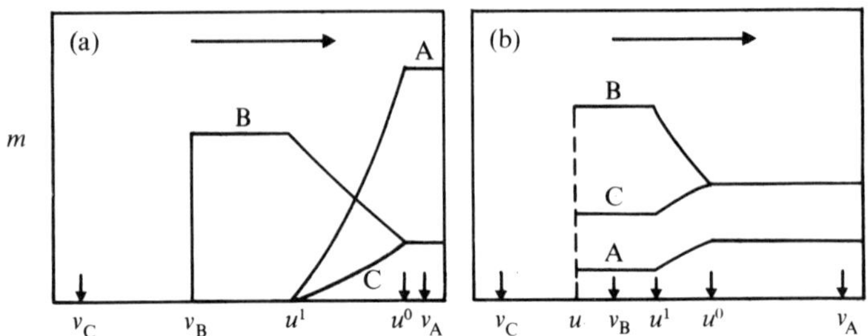

FIG. 4.4. Migration patterns for the case $v_A > v_B > v_C$, showing two possible solutions. Each pattern may be regarded as representing *either* a descending side, with slower reactant labelled B, *or* an ascending side, with slower reactant labelled A. (Adapted from Gilbert and Jenkins (1959).)

$v_A > v_C > v_B$

As Gilbert and Jenkins (1959) have noted, eqns (4.27d), (4.28), and (4.29) describe the reaction boundary region for this case. However, in contrast to the situation with $v_C > v_A > v_B$ (where λ from eqn (4.22d) was greater than unity), $-1 < \lambda < 1$ for the case under discussion, and, consequently, $(\lambda - 1)$ is necessarily a negative quantity. It follows from inspection of eqn (4.28) that Φ can only be positive, since m_A must be positive, and that eqns (4.30a, b) but not eqns (4.31a, b) therefore hold. The implication is that a plateau region of B (but not of A) may exist, terminated on one side at $v_B t$ by the solvent plateau and on the other side at $u't$ by the reaction boundary. This is true on both ascending and descending sides, but the patterns are nonetheless non-enantiographic, since the slower reactant is labelled B on the descending side but B refers to the faster-moving reactant on the ascending side. This situation is clearly illustrated by the numerical examples of Gilbert and Jenkins (1959), both for a symmetrical situation ($m_A/k = m_B/k$; their Fig. 2) and for an unsymmetrical case (their Fig. 3). The latter example also shows that the reaction boundary on the ascending side may appear bimodal in a simulated schlieren trace. Practical illustrations of these observations have been provided by moving boundary electrophoresis of the systems pepsin + bovine serum albumin (Cann and Klapper 1961) and yeast RNA + ovomucoid (Longsworth 1959*a*). Fig. 4.5 presents results obtained by Longsworth: the bimodality of the reaction boundary on the ascending side is clearly evident at pH 4·63 when an excess of the slower moving reactant, ovomucoid, is present in the original mixture. However, it is noted that in a different environment (pH 3·92) and with nucleic acid in excess, the reaction boundary on the descending side appears markedly bimodal. Similar complicated migration patterns have been obtained in moving boundary electrophoresis of mixtures of insulin and protamine (Timasheff and Kirkwood 1953), pepsin and serum albumin (Cann and Klapper 1961), and antigen and antibody (Singer *et al.* 1959). Gilbert and Gilbert (1965) have extended the treatment to permit calculation of migration patterns for systems of the type $A+B \rightleftharpoons C_1$, $B+C_1 \rightleftharpoons C_2$, which must occur in the reaction of a univalent antigen with a bivalent antibody, the situation studied by Singer *et al.* (1959).

Investigators are cautioned against facile quantitative interpretation of such

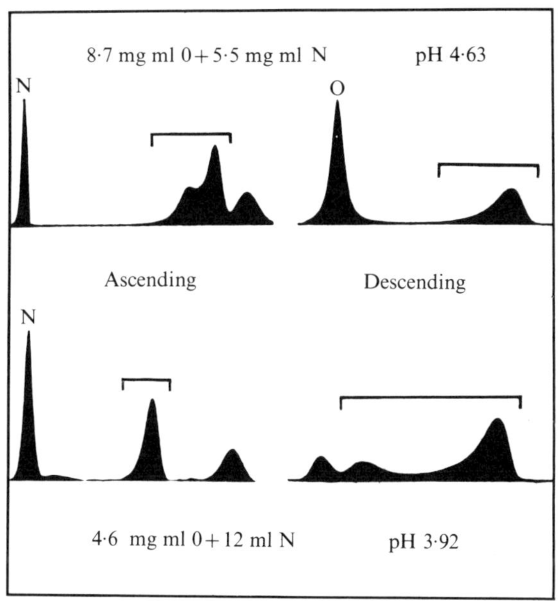

FIG. 4.5. Moving boundary electrophoretic patterns of interacting mixtures of ovomucoid, 0, and yeast nucleic acid, N, in 0·1 *I* acetate buffers: reaction boundaries are denoted by square brackets. (Adapted from Longsworth (1959*a*).)

patterns. In particular, the changes in concentration across the boundaries representing pure reactants on either the ascending or descending sides do not equal their respective equilibrium concentrations in the original mixture. Indeed, the numerical examples of Gilbert and Jenkins (1959) suggest that the concentrations of pure reactants may be expected to be larger than the corresponding equilibrium values (cf. the smaller concentrations predicted for $v_C > v_A > v_B$). It may therefore be concluded that with systems of the type $A+B \rightleftharpoons C$ the concentration of a pure, separated reactant may be used directly in eqn (4.18) to compute the equilibrium constant of the reaction only in the event that the velocity of the complex is identical with that of one of the reactants.

(c) *A note on the validity of simultaneous continuity equations*

Prediction of the nature of migration patterns in this chapter has followed the approaches of Gilbert and Jenkins (1959) and Nichol and Ogston (1965*a*). Fundamentally, both approaches are based on eqns (2.11a, b) which follow intuitively from the derivation of eqn (2.5), the continuity equation for the migration of a single solute. It does not concern us here that eqn (2.5) assumes implicitly the neglect of diffusional flows, a rectangular cell, etc. (these important aspects being treated separately in Chapter 5); but rather, that eqn (2.5) postulates the existence of a lamina of *both* constant velocity and concentration. As Nichol *et al.* (1970) have argued, this is a justifiable basis for the formation of a continuity equation describing the migration of a single solute, whether it be non-self-interacting (eqn (2.5)) or self-interacting (eqn (2.8)). Indeed, a geometrical interpretation of these equations is that the loci of constant composition within the boundary are straight lines on the x, t plane converging on the origin, and thus the boundary consists of a continuum of regions of constant composition, each moving with constant velocity. As a consequence, there are no inconsistencies in the use of eqn (2.8) in the prediction of general boundary forms for the migration of a system comprising one constituent (for example, a polymerizing solute): that is, the findings in Chapter 3 for systems $n\mathrm{A} \rightleftharpoons \mathrm{C}$ are valid, being based on the rigorously derived eqn (2.8). However, in the use of the corresponding set of continuity equations (eqns (2.11a, b)) for a system comprising two constituents, we have consistently assumed not only that laminae of constant velocity and composition exist, but also that u_{A} and u_{B} are identical between these equations. For example, in writing eqn (4.6) or eqn (4.22b) we have treated eqns (4.2a, b) as if they were simultaneous equations; but these latter equations follow from eqns (2.11a, b) only by assuming that $u_{\mathrm{A}} \equiv u_{\mathrm{B}}$. There is in fact no general justification for this latter assumption.

Nichol *et al.* (1970) have viewed the problem of the migration of a multisolute system differently. They defined the ith constituent concentration, $m_i(x, t)$, as the sum, $\sum m_j$, where m_j is the concentration of the jth molecular species and the sum is extended to all species containing the ith constituent. The ith constituent velocity was termed $v_i(x, t)$ and written in a summed form as illustrated in eqns (2.6b) and (2.9c, d). In these terms eqn (2.1) becomes

$$\frac{\partial m_i}{\partial t} + \frac{\partial (v_i m_i)}{\partial x} = 0 \qquad (i = 1, 2, \ldots n),$$

where n is the number of constituents: this equation necessarily applies to each constituent, since it stipulates the condition of conservation of mass (neglecting diffusion, etc., as before). This equation may be written as

$$\frac{\partial m_i}{\partial t} + \sum_{j=1}^{n} a_{i,j} \frac{\partial m_j}{\partial x} = 0 \qquad (i = 1, 2, \ldots n), \tag{4.33a}$$

where

$$a_{i,j} = \frac{(\partial v_i m_i)}{\partial m_j} \qquad (i, j = 1, 2, \ldots n). \tag{4.33b}$$

Nichol et al. (1970) classified boundary types as hyperbolic, parabolic, and elliptic, on the basis of the nature of the roots of the characteristic equation of the determinant, D_n, formed by the derivatives, $a_{i,j}$. One-constituent systems (for example, $n\mathrm{A} \rightleftharpoons \mathrm{C}$) are of the hyperbolic type, to which eqn (2.8) applies. In a special case of a system comprising more than one constituent it is possible that all elements of a row or of a column in D_n may be zero, whence D_n may be written as the product of two analogous determinants of order $(n-1)$ and 1, respectively: the system can then be considered as a combination of two independent systems of $(n-1)$ and 1 constituents, respectively. This occurs when $\partial v_i/\partial m_j = 0$ for a given i and any $j \neq i$, or for a given j and any $i \neq j$. Examples are provided by the systems $v_\mathrm{A} = v_\mathrm{C} > v_\mathrm{B}$ and $v_\mathrm{A} > v_\mathrm{B} = v_\mathrm{C}$, where it follows from eqns (2.9c, d) (assuming velocities to be independent of composition) that $\partial \bar{v}_\mathrm{A}/\partial m_j = 0$ and $\partial \bar{v}_\mathrm{B}/\partial m_j = 0$, respectively. Since these systems comprise two constituents (and this remains true whether or not the acceptor and ligand are self-interacting), they may be considered as two independent systems, each of one constituent. Accordingly, the use of eqns (2.11a, b) or (2.17a, b) in Chapter 2 for systems of this type may be judged satisfactory. Figure 4.1 illustrates (but does not itself prove) the identity of the median bisectors of the gradients of constituent concentrations on both ascending and descending sides: this follows as a consequence of equating u_A and u_B, a reasonable identification for the systems under discussion.

More generally, Rescigno (personal communication) has pointed out that for a two-constituent system the equation $D_n = 0$ has two separate real roots or a double root when, and only when,

$$\left(\bar{v}_\mathrm{A} - \bar{v}_\mathrm{B} + \bar{m}_\mathrm{A} \frac{\partial \bar{v}_\mathrm{A}}{\partial \bar{m}_\mathrm{A}} - \bar{m}_\mathrm{B} \frac{\partial \bar{v}_\mathrm{B}}{\partial \bar{m}_\mathrm{B}}\right)^2 + 4\bar{m}_\mathrm{A}\bar{m}_\mathrm{B} \frac{\partial \bar{v}_\mathrm{A}}{\partial \bar{m}_\mathrm{A}} \frac{\partial \bar{v}_\mathrm{B}}{\partial \bar{m}_\mathrm{B}} \geq 0.$$

This condition dictates whether or not eqns (2.11a, b) may be used to formulate migration patterns. As we have seen, the condition is clearly fulfilled for systems in which $v_\mathrm{A} = v_\mathrm{C} > v_\mathrm{B}$ and $v_\mathrm{A} > v_\mathrm{B} = v_\mathrm{C}$, and it is also valid for those involving reactants with equal velocities ($v_\mathrm{A} = v_\mathrm{B}$): thus the solution shown in Fig. 4.2 is also soundly based. It is gratifying that the solutions shown in Figs. 4.1 and 4.2 were, indeed, the ones used for the determination of equilibrium constants for systems of the type $A + B \rightleftharpoons C$. The situation pertaining to cases involving no equalities between individual species velocities requires further clarification. From experimental results it seems likely that the general forms of the boundary distributions are as described: certainly, the available solutions of Gilbert and Jenkins warn against facile quantitative interpretation of boundary forms and positions, and against the direct use of concentrations of separated reactants for evaluating equilibrium constants. Nevertheless, it is relevant that the solutions obtained for these cases are based on eqns (2.11a, b) and their validity in this connection has not been established.

4.2. Protein interactions with buffer constituents

Since most mass migration experiments performed with protein preparations are conducted in a buffered environment, it is timely to consider some effects of rapidly-equilibrating heterogeneous association reactions between the protein and constituents of the buffer. The concept that protons may be regarded as reactants in this sense is familiar (for example, Longsworth 1959*a*); but migration experiments conducted with a variety of different proteins (Cann 1958, 1959, 1960, 1961; Cann and Phelps 1957, 1959; Chiancone and Gilbert 1965; Phelps and Cann 1956) have also indicated that undissociated buffer acid (acetic and other carboxylic acids) may also function in this way. Cann and Goad (1965*a*, *b*) visualized an overall equilibrium reaction $A + n\mathrm{HR} \rightleftharpoons A(\mathrm{HR})_n$, where A denotes protein and HR the buffer acid. They suggest that the complex, $A(\mathrm{HR})_n$, could be characterized by a different sedimentation coefficient from that of A due to a conformational change induced by the binding, or by a different electrophoretic mobility due perhaps primarily to a change in the net charge caused by a difference between pK values of carboxyl groups on free and complexed protein. Regardless of the detailed reasons for the possible differences between the velocity terms, it is important to assess the effects that such an equilibrium involving species with different velocities would have on migration patterns.

Cann and Goad (1965*a*) wrote two continuity equations in terms of the *constituent* concentrations of A and HR. These equations are formally similar to eqns (2.11a, b), but are non-linear because second-order differential terms are included to account for diffusional flows: examples of such equations will be given in Chapter 5 when the effects of diffusion are considered in greater detail. It suffices here to note that finite difference approximations to the continuity equations were written and used to calculate numerical examples.

Methods are detailed by Goad in Cann (1970). The particular solutions required initial boundary conditions, and two sets were chosen: the first corresponded, as in moving boundary electrophoresis, to a sharp boundary between the original solution containing A, HR, and $A(HR)_n$ and the solvent containing the same concentration of HR (Cann and Goad 1965*a*); the second (Cann and Goad 1965*b*) corresponded to a sharp zone of the original solution embedded in a supporting medium containing the same concentration of HR.

For the examples simulated to preserve a plateau region of original solution, patterns (plots of gradients in constituent concentration of A vs. a distance parameter, together with the distribution of HR) were computed for conjugate ascending and descending sides: a range of values of n, time, and the equilibrium ratio of A to $A(HR)_n$ in the original solution was considered. If the relative concentration of HR was sufficiently large (low n or low constituent protein concentration), a single boundary was predicted, migrating with the average mobility of A and $A(HR)_n$. Under other conditions enantiographic patterns with two boundaries on each side were found, falsely suggesting that two non-interacting protein solutes were present. In some examples bimodal reaction boundaries appeared on each side, there being a tendency to sharpen on one side and to spread on the conjugate side. Since it was assumed in all calculations that the mobility of $A(HR)_n$ was constant and greater than that of A, it is not surprising that certain numerical examples bear similarity with those presented by Gilbert and Jenkins (1959) for the conjugate cases $v_C > v_A > v_B$ and $v_C > v_B > v_A$ (B = HR, C = $A(HR)_n$).

In the simulation of zonal migration experiments, Cann and Goad (1965*b*) presented their numerical examples as plots of constituent concentration of A against distance, together with the concentration gradient in HR produced. One of their interesting results was that two bands (or zones) may appear (although the constituent concentration of A never fell to zero between them), and that these zones may either remain apparently resolved or, with increasing time, form a broad single zone with a trailing edge. A striking practical example of such behaviour was observed by Cann (1966): a control zonal electrophoretic experiment on bovine serum albumin in phosphate buffer (0·0043 M Na_2HPO_4, 0·030 M NaH_2PO_4, pH 6·1) yielded a single band on a cellulose acetate strip, whereas electrophoresis of the same initial concentration of the protein in 0·012 M Na_2HPO_4, 0·008 M NaH_2PO_4, 0·4 M H_3BO_3(pH 6·1, similar ionic strength) resulted in the detection of two bands by the same staining procedure. Cann employed two methods to show that the *apparently* 'better resolution' in the phosphate–borate medium could not be attributed to microheterogeneity of the protein. First, samples from the two zones were obtained by eluting unstained strips with the buffer, and re-subjected to electrophoresis in the same buffer: each zone again gave rise to two bands. Secondly, moving boundary electrophoresis of bovine serum albumin in the phosphate–borate buffer yielded nonenantiographic ascending and descending patterns: the reader will recognize from the contents of this chapter that such behaviour is a general property of migration patterns obtained with rapidly-equilibrating systems.

The important general implication of the findings of Cann and co-workers is that the appearance of multiple bands in zonal experiments need not necessarily be due to heterogeneity. Bethune and Kegeles (1961*a*, *b*) also stressed that rapid chemical re-equilibration could lead to a reaction zone which is bi- or tri-modal. With a relatively insensitive staining procedure, bands (corresponding to the maxima) may appear to separate, although in fact all solute species coexist in equilibrium in all regions of the zone. As a consequence, if a sample were taken from any part of the reaction zone, concentrated to the initial value, dialysed against the original buffer, and re-run, the resultant pattern would be the same as that obtained initially. This would also be the situation if a complicated reaction zone reflected kinetically controlled isomerization or polymerization of the protein (see section 5.1 and Fig. 5.1(b)). On the other hand, if the resolution into bands were due solely to heterogeneity, the pattern obtained on rerunning would indicate an enrichment of certain components in the excised band. An investigator who is confronted with proof that some type of chemical equilibration is complicating zonal migration experiments is advised to characterize the system further with the use of migration experiments involving a plateau region of original solution composition, or equilibrium experiments, such as sedimentation equilibrium and light scattering.

4..3. Hybridization and isomerization

The use of continuity equations to predict the nature of migration patterns is not restricted to systems undergoing polymerization reactions or heterogeneous associations. Two other types of equilibria, which were introduced in Chapter 1 as being of biological importance, are rapidly re-equilibrating hybridization and isomerization reactions. The behaviour of systems of these types in mass migration experiments involving a plateau region of original solution composition may be deduced from the appropriate continuity equations in a manner entirely analogous to that adopted in the treatments of the other systems. If the hybridization reaction $A+B \rightleftharpoons C+D$ (i.e. $p_1 p_2 + q_1 q_2 \rightleftharpoons p_1 q_1 + p_2 q_2$) is considered, and the convention adopted that $v_A \geqq v_B$, $v_C \geqq v_D$ with distances positive in the direction solvent→solution for both the ascending and descending sides, there are fourteen cases involving different relative magnitudes of individual species velocities (Nichol and Ogston 1967). Each case was discussed, and it suffices for the purpose of illustration to examine in detail only one of the solutions. We select the situation $v_A = v_B > v_C = v_D$, because under special circumstances the solution also describes the behaviour of an isomerizing system, and thereby provides a link with the material given in Chapter 3.

$v_A = v_B > v_C = v_D$

The required basic equations have been given as eqns (2.12)–(2.14), inclusive, which may be reformulated as follows. Differentiation of the equilibrium condition (eqn (2.13)) yields

$$\frac{dm_A}{m_A} + \frac{dm_B}{m_B} = \frac{dm_C}{m_C} + \frac{dm_D}{m_D}, \tag{4.34}$$

while differentiation of eqns (2.12a, d) and substitution into eqn (2.14) gives

$$(u - v_A)\, dm_A + (u - v_C)\, dm_C = 0, \tag{4.35a}$$

$$(u - v_B)\, dm_B + (u - v_C)\, dm_C = 0, \tag{4.35b}$$

where all u have again been assumed to be identical: this is necessarily true when a sharp or hypersharp boundary is encountered, since there can only be one median bisector of such boundaries. For example, if a hypersharp boundary separates two plateaux, α ($m_A = m_A^\alpha$, $m_B = m_B^\alpha$, $m_C = m_C^\alpha$, $m_D = m_D^\alpha$) and β ($m_A = m_A^\beta$, $m_B = m_B^\beta$, $m_C = m_C^\beta$, $m_D = m_D^\beta$), the unique position of the hypersharp boundary is $\bar{u}t$, where $\bar{u}$ is given by

$$\bar{u} = \frac{\int_\beta^\alpha u\, d\bar{m}_Y}{\int_\beta^\alpha d\bar{m}_Y}, \tag{4.36a}$$

which may be rewritten, using eqn (2.14), as

$$\bar{u} \int_\beta^\alpha d\bar{m}_Y - \int_\beta^\alpha d\bar{v}_Y \bar{m}_Y = 0. \tag{4.36b}$$

Performing the integrations between the specified limits, and rearranging with the aid of eqns (2.12a, b, c) leads to the following independent equations:

$$(\bar{u} - v_A)(m_A^\alpha - m_A^\beta) + (\bar{u} - v_C)(m_C^\alpha - m_C^\beta) = 0, \tag{4.37a}$$

$$(\bar{u} - v_A)(m_A^\alpha - m_A^\beta) + (\bar{u} - v_D)(m_D^\alpha - m_D^\beta) = 0, \tag{4.37b}$$

$$(\bar{u} - v_B)(m_B^\alpha - m_B^\beta) + (\bar{u} - v_C)(m_C^\alpha - m_C^\beta) = 0. \tag{4.37c}$$

It is now possible to proceed with the discussion of the solution, which is illustrated in Fig. 4.6.

(i) When $u > v_A = v_B$, it is required that the original solution plateau extend, wherein $m_A = m_{o,A}$, $m_B = m_{o,B}$, $m_C = m_{o,C}$, and $m_D = m_{o,D}$.

(ii) When $u = v_A = v_B$, it is clear that $(u - v_C) = (u - v_D) \neq 0$, while eqns (4.35a, b) show that $dm_C = 0$ (i.e. $m_C = m_{o,C}$) and $dm_D = 0$ ($m_D = m_{o,D}$). It follows from eqn (4.34) that

$$\frac{dm_A}{dm_B} = \frac{-m_A}{m_B} < 0 \qquad \text{(provided } dm_A \neq 0,\ dm_B \neq 0\text{)} \tag{4.38a}$$

or

$$dm_A m_B = 0. \tag{4.38b}$$

Equation (4.38a) implies that if $m_{o,A} > m_{o,B}$, then m_A decreases across the sharp boundary at $ut = v_A t$ and m_B increases, the situation being reversed if $m_{o,B} > m_{o,A}$. Integration of eqn (4.38b) yields

$$m'_A m'_B = m_{o,A} m_{o,B} \equiv \Pi, \tag{4.38c}$$

which shows that neither m'_A nor m'_B is zero above the sharp boundary at $v_A t$, which therefore separates the original solution plateau from one wherein $m_A = m'_A$, $m_B = m'_B$, $m_C = m_{o,C}$, and $m_D = m_{o,D}$. The values of m'_A and m'_B remain to be specified.

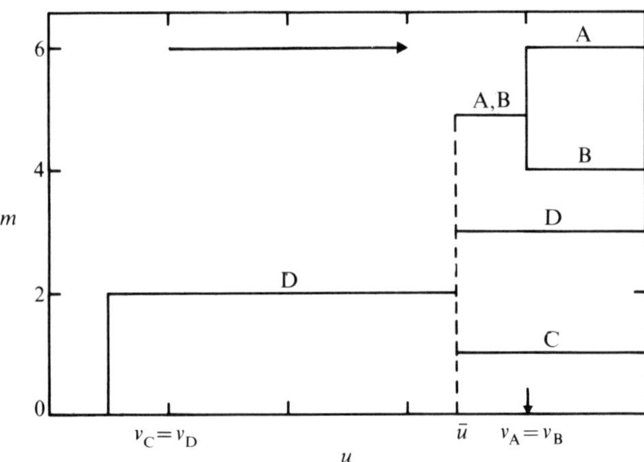

FIG. 4.6. Numerical example of a representative migration pattern for the case $v_A = v_B > v_C = v_D$, the broken line indicating a hypersharp boundary. This pattern refers *either* to a descending side with C and D as the slower-migrating species, *or* to the conjugate ascending side with the appropriate relabelling. Values of the various parameters (in arbitrary units) are as follows: $m_{o,A} = 6$, $m_{o,B} = 4$, $m_{o,C} = 1$, $m_{o,D} = 3$, $v_A = v_B = 8$, $v_C = v_D = 1$. (Adapted from Nichol and Ogston (1967).)

(iii) When $u = v_C = v_D$, eqns (4.35a, b) insist that $dm_A = 0$ (i.e. $m_A = 0$) and $dm_B = 0$ ($m_B = 0$), since the solvent plateau must extend when $u < v_C = v_D$. Equation (4.34) does not permit dm_C/dm_D to be negative, because this would imply that m_C or m_D become negative; but at $v_C t$ eqns (2.13) and (4.35a, b) do permit *either* $dm_C \neq 0$ *or* $dm_D \neq 0$. For the purpose of illustration we choose the latter situation, in which m_D rises from zero to a plateau of m''_D. In this case, a boundary in the region $v_A = v_B > u > v_C = v_D$ must separate the plateaux ($m_A = m_B = m_C = 0$, $m_D = m''_D$) and ($m_A = m'_A$, $m_B = m'_B$, $m_C = m_{o,C}$, $m_D = m_{o,D}$).

(iv) It may be shown (Nichol and Ogston 1967) that the latter boundary cannot be spread, and it therefore appears that a hypersharp boundary occurs at $\bar{u}t$, where $v_A = v_B > \bar{u} > v_C = v_D$. In this case the value of $\bar{u}$ is uniquely defined by eqns (4.37a–c), which become

$$(\bar{u} - v_A)m'_A + (\bar{u} - v_C)m_{o,C} = 0, \tag{4.39a}$$

$$(\bar{u} - v_A)m'_A + (\bar{u} - v_D)(m_{o,D} - m''_D) = 0, \tag{4.39b}$$

$$(\bar{u} - v_B)m'_B + (\bar{u} - v_C)m_{o,C} = 0. \tag{4.39c}$$

Subtracting eqn (4.39c) from eqn (4.39a), and dividing by $(\bar{u} - v_A) = (\bar{u} - v_B) \neq 0$ gives

$$m'_A = m'_B = \Pi^{1/2}. \tag{4.40}$$

Thus across the sharp boundary at $v_A t$ we find that either m_B decreases and m_A increases until $m'_A = m'_B$ (in the event that $m_{o,B} > m_{o,A}$), or m_B increases and m_A decreases (if $m_{o,A} > m_{o,B}$). Subtracting eqn (4.39b) from eqn (4.39a) and dividing by $(\bar{u} - v_C) = (\bar{u} - v_D) \neq 0$ gives

$$m''_D = m_{o,D} - m_{o,C}, \tag{4.41}$$

which shows that a boundary in D will arise at $v_D t$ if $m_{o,D} > m_{o,C}$, or a boundary in C will arise in the event that $m_{o,C} > m_{o,D}$ (whereupon $m''_C = m_{o,C} - m_{o,D}$). The position $\bar{u}t$ may now be found in terms of known quantities. For $m_{o,D} > m_{o,C}$, eqn (4.39a) yields

$$\bar{u} = \frac{v_A \Pi^{1/2} + v_C m_{o,C}}{\Pi^{1/2} + m_{o,C}}, \tag{4.42}$$

whereas for $m_{o,C} > m_{o,D}$ it may be shown that

$$\bar{u} = \frac{v_A \Pi^{1/2} + v_D m_{o,D}}{\Pi^{1/2} + m_{o,D}}. \tag{4.43}$$

With systems of this type it is therefore possible that three boundaries may arise on ascending or descending sides. In practice, diffusional flows would tend to spread the sharp boundaries at $v_A t$ and $v_C t$, but the boundary between them may well remain hypersharp. If v_A differed very little from v_C this could lead to a composite schlieren peak, spread on both sides around a central sharp 'spike'. Much more careful investigation is required before unusual schlieren patterns, sometimes observed in practice, could be matched with this solution or others provided by Nichol and Ogston (for example, $v_A = v_B = v_C > v_D$, for which two sharp boundaries are predicted, or the conjugate case $v_C > v_A = v_B = v_D$, where a sharp boundary is adjacent to a spread boundary). Moreover, Nichol and Ogston (1967) did not provide solutions for situations in which all individual species velocities were unequal: they suggested that it would be possible to indicate the locations of plateaux and boundaries, but that the exact character and limits of some of the latter would remain unspecified unless unwarranted assumptions were made concerning the existence of laminae of constant composition and velocity.

The special case of isomerization reactions. The behaviour of a rapidly-equilibrating isomerizing system ($n\text{A} \rightleftharpoons \text{C}, n=1$) in mass-migration processes may be deduced in two ways, which are in fact equivalent. First, we may again consider the case $v_A = v_B > v_C = v_D$, with the additional restrictions that $m_{o,A} = m_{o,D}$ and $m_{o,C} = m_{o,D}$. Equation (4.41) suffices to show that no boundary in either C or D will arise at $ut = v_C t = v_D t$. Similarly, it follows from eqns (4.38c) and (4.40) that $m'_A = m'_B = m_{o,A} = m_{o,B} = \Pi^{1/2}$, and thus no change in the concentration of any species occurs at $ut = v_A t = v_B t$. Only a single boundary is therefore observed on both ascending and descending sides, its velocity, $\bar{u}$, being given by eqn (4.44).

$$\bar{u} = \frac{v_A m_{o,A} + v_C m_{o,C}}{m_{o,A} + m_{o,C}}. \tag{4.44}$$

Even though the concentrations of A and C have been expressed on a molar scale, $\bar{u}$ as defined by eqn (4.44) is clearly the weight-average velocity of an isomerizing system, since $M_A = M_C$.

Secondly, as noted in Chapter 3, we could approach the problem by placing $n=1$ in eqn (3.2), whereupon eqn (4.44) results after noting that $K = c_{o,C}/c_{o,A} = m_{o,C}/m_{o,A}$. Again a single boundary moving with the weight-average velocity is predicted. Determination of $\bar{u}$ would permit evaluation of the isomerization constant, K, provided values of v_A and v_C were available, from eqn (4.45),

$$K = \frac{v_A - \bar{u}}{\bar{u} - v_C}, \tag{4.45}$$

which follows directly from rearrangement of eqn (4.44), or from substitution of $n = 1$ in eqn (3.8). Of course, it may be impossible to obtain estimates of v_A and v_C: indeed, the few examples of reported isomerization constants (Changeux and Rubin 1968, Nichol *et al.* 1969*c*) have followed from treatments of binding results (Chapter 1) rather than from interpretation of mass migration results.

5

FRICTIONAL EFFECTS AND OTHER COMPLICATING FACTORS

IN the preceding three chapters the treatment of the general forms of boundary distributions that may arise in mass migration experiments has been restricted to chemically reacting systems for which the dependence of the relevant constituent velocities on composition is described solely by the law of mass action. Moreover, only rapidly equilibrating systems have been considered, where all constituent velocities are independent of time. It is clear that not all systems studied in practice will conform with these restrictions placed on the functional relations of constituent velocities: one aim of the present chapter is to extend the treatment to other types of interacting system. At the same time, it is important to note also that, in a practical situation, various assumptions made in deriving the basic continuity equations (for example, eqns (2.5) and (2.8)) will only be imperfectly fulfilled. These assumptions will therefore be examined further.

5.1. Slow equilibria and those subject to kinetic control

(a) *Frontal experiments*

For a system comprising one constituent (for example, an isomerizing or a polymerizing solute), which is equilibrating so rapidly that the local equilibrium between the states of the solute molecules is instantaneously established at each point in the cell, the constituent velocity, $\bar{v}$, is independent of time, and is therefore only a function of the constituent concentration, $\bar{c}$. It follows from eqn (2.8) that u is also only a function of $\bar{c}$, and hence that the *form* of the migration pattern ($\bar{c}$ vs. u, or $\mathrm{d}\bar{c}/\mathrm{d}u$ vs. u) is independent of time. Plots of $\bar{c}$ vs. x or $\mathrm{d}\bar{c}/\mathrm{d}x$ vs. x may readily be obtained for any selected time t, since x, the distance from the origin, equals ut. However, with systems comprising one or more constituents in which constituent velocities are functions of time, no such time-independence of the pattern form can pertain.

One example of a system in which the constituent velocity is a function of time is provided by the mass migration of an isomerizing or a polymerizing solute ($n\mathrm{A} \rightleftharpoons \mathrm{C}$) under conditions where the rate of interconversion between states A and C is *comparable* with the rate at which they *tend* to separate by differential migration ($v_\mathrm{A} \neq v_\mathrm{C}$). This has been termed the kinetically controlled case (Longsworth and MacInnes 1942, Nichol *et al.* 1964), because the continuity equations must necessarily include terms for the rates of appearance and disappearance of material due to chemical reaction, as well as terms for directed migration. Thus Van Holde (1962) and Belford and Belford (1962) formulated the continuity equation in terms of weight concentrations as

$$\left(\frac{\partial c_{\mathrm{A}}}{\partial t}\right)_x = -v_{\mathrm{A}}\left(\frac{\partial c_{\mathrm{A}}}{\partial x}\right)_t - k_{+1}c_{\mathrm{A}}^n + k_{-1}c_{\mathrm{C}},$$

$$\left(\frac{\partial c_{\mathrm{C}}}{\partial t}\right)_x = -v_{\mathrm{C}}\left(\frac{\partial c_{\mathrm{C}}}{\partial x}\right)_t + k_{+1}c_{\mathrm{A}}^n - k_{-1}c_{\mathrm{C}}, \qquad (5.1)$$

where k_{+1} and k_{-1} are rate constants pertaining to the forward and reverse reactions, respectively. It is noted that the species velocities v_{A} and v_{C} are considered to be constants (i.e. frictional interactions continue to be ignored) and that there are no second-order differential terms to account for diffusional flows. Van Holde (1962) considered an isomerizing system ($n=1$), in which case it was possible to employ Laplace transforms to obtain an analytical solution describing the migration patterns: the initial boundary condition specified a sharp boundary between solvent and a plateau of the original equilibrium mixture. Belford and Belford (1962) examined the case where $n=2$ (a dimerizing system), and pointed out that the non-linearity introduced into the reaction terms of eqn (5.1) prevented the use of Laplace transforms to obtain analytical solutions: however, by numerical methods they illustrated particular examples of migration patterns. Numerical solutions for migration of a kinetically controlled isomerizing system ($n=1$) have also been provided by Cann, Kirkwood, and Brown (1957); Cann and Bailey (1961); and Scholten (1961). These workers included a second order differential term in the continuity equation to account for the diffusional flows of A and C in terms of concentration-independent diffusion coefficients D_{A} and D_{C}. The relevant terms for addition to the right-hand sides of eqn (5.1) are $D_{\mathrm{A}}(\partial^2 c_{\mathrm{A}}/\partial x^2)_t$ and $D_{\mathrm{C}}(\partial^2 c_{\mathrm{C}}/\partial x^2)_t$, respectively: these follow from a combination of Fick's first and second laws of diffusion, together with an assumed absence of coupled flows. Inspection of the numerical examples provided by all of these workers leads to the dramatic conclusion that for either isomerization or dimerization, plots of $\mathrm{d}\bar{c}/\mathrm{d}x$ vs. x may exhibit one, two, or three peaks, depending on the values selected for the various parameters and for the time: an example is shown in Fig. 5.1(a). In common with rapid reactions, peaks observed with a system subject to kinetic control cannot be identified with individual species, but in contrast with the mass migration behaviour of rapidly interacting one-constituent systems, the number of observed peaks may exceed the number of individual solute species.

The experimenter has control over some of the parameters which (as the numerical examples show) influence the migration patterns obtained with kinetically controlled systems. For example, the nature of the applied field (electrical or gravitational) governs the difference in magnitude and possibly sign of the velocities v_A and v_C, but presumably has little effect on the rate of re-equilibration between A and C. A change in magnitude of a particular applied field will also affect the rate at which the species tend to separate by differential migration, whereas a change in temperature may operate primarily by influencing the rate (and possibly position) of re-equilibration. Thus it is conceivable that a one-constituent system studied under one set of conditions may exhibit the migration behaviour of a kinetically controlled system, while under another, the behaviour of a rapidly interacting system (Chapter 3). In this sense the classification of systems as rapidly interacting or kinetically controlled is arbitrary.

In Chapter 3 we have seen that the effect of varying $\bar{c}$ on the observed migration patterns was extremely useful in the characterization of rapidly interacting systems comprising a single constituent, and the question arises whether variation of $\bar{c}$ also affects patterns obtained with kinetically controlled systems. The numerical examples provided by Oberhauser, Bethune,

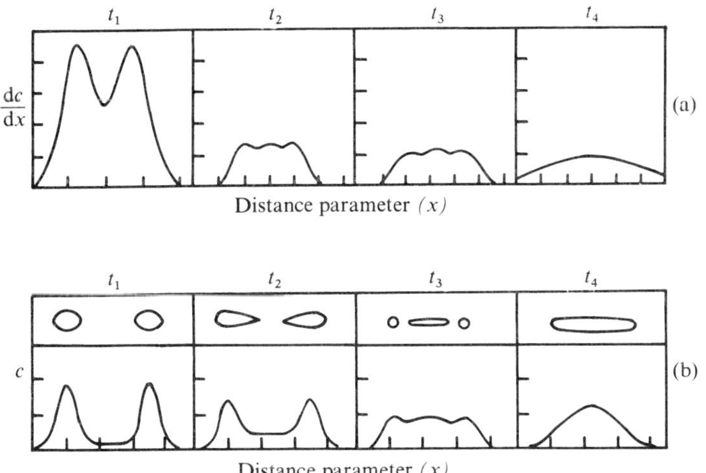

FIG. 5.1. Numerical examples of migration patterns obtained with isomerizing systems subject to kinetic control. (a) Schlieren patterns predicted on the basis of a symmetrical choice of relevant parameters, the experiment involving a plateau region of original composition throughout its duration: $t_1 < t_2 < t_3 < t_4$. (Adapted from Cann and Bailey (1961).) (b) Concentration distributions (lower sections) and possible spot outlines after staining (upper sections) in a zonal migration experiment: $t_1 < t_2 < t_3 < t_4$. (Adapted from Keller and Giddings (1960).)

and Kegeles (1965), which relate to a dimerizing system, clearly illustrate that holding all other parameters (including time) constant, and varying $\bar{c}$ influences the pattern to the extent that the number of peaks may vary. One disadvantage of most of the treatments of kinetically controlled systems is the absence of any information on the nature of the conjugate side observable in migration experiments involving a plateau of original composition. Since non-enantiography of conjugate sides has been established for rapid interactions (isomerizations excepted), it would be of interest to explore the behaviour of kinetically controlled systems in this regard: determination of the conjugate pattern should present no great problems in computation.

(b) *Zonal experiments*

It is of interest to compare the behaviour of kinetically controlled systems in mass migration experiments which lack a plateau of original composition, with that in experiments which preserve such a region (Fig. 5.1(a) and (b)). These numerical examples (Cann and Bailey 1961, Keller and Giddings 1960), both pertaining to an isomerizing system, were calculated with different sets of parameters (although in each case the parameters were chosen symmetrically, for example, $k_{+1} = k_{-1}$): however, the similarity between the general behaviour is striking. In Fig. 5.1(b) it is seen that as time progresses, the initial zone may develop into two spots corresponding to maxima in the concentration profile, with a diffuse streak between them if the assay procedure were sufficiently sensitive. As time increases, the following effects may be observed: (i) the two intense bands may become asymmetric (t_2 in Fig. 5.1(b)); (ii) three bands may be observed superimposed on a blurred background (t_3 in Fig. 5.1(b)); (iii) a blurred band may develop (t_4). We note also that in Fig. 5.1(a) one, two, or three maxima may be observed in the concentration gradient patterns from frontal experiments, and that the gradient never decreases to zero between maxima. The final symmetrical distributions in Fig. 5.1(a) and (b) result from the inclusion of diffusional spreading in the two sets of calculations. Of major importance is the fact that in both types of migration experiment *reaction* boundaries or *reaction* zones are formed which cannot be interpreted readily in terms of properties relevant to the individual solute species. By sampling any part of the reaction boundary or reaction zone and re-running the sample after reconcentrating and dialysing to the original condition, it is possible to distinguish between kinetically controlled systems, and those comprising a heterogeneous mixture of non-interacting solutes. Variation of diverse parameters such as the magnitude of the applied field, temperature, and constituent concentration also provide valuable diagnostic tests: but it appears that further elucidation of kinetically controlled systems (especially those comprising more than one constituent) would better be performed by the equilibrium methods of light scattering or sedimentation equilibrium.

(c) *Systems undergoing slow re-equilibration*

The classification of reactions as rapidly-equilibrating and kinetically controlled must logically be extended to encompass those reactions where equilibration is so slow, compared with the rate of differential migration of the solute species, that no noticeable effects of equilibration are observed in the migration experiment. Interpretation of migration patterns obtained with such systems offers no difficulty: the concentration gradient profile from a plateau experiment exhibits discrete peaks, each of which represents a gradient of concentration due to an individual species (provided that each species is characterized by a different velocity and that sufficient time for resolution is allowed). An example is presented in Fig. 2.1(b) and (c), where A and B may now be considered as two slowly equilibrating isomers or polymers. The observed velocities, whether sedimentation coefficients, electrophoretic mobilities or elution volumes, may be related to properties of the individual species such as size, shape, hydration, and net charge (Svedberg and Pedersen 1940; Schachman 1959; Abramson, Moyer, and Gorin 1942; Ackers 1964; Andrews 1966). The area under each peak may be related to the equilibrium concentration of the relevant species in the original solution plateau, and may therefore be used to evaluate the apparent equilibrium constant. It is also clear from Fig. 2.1(b) and (c) that ascending and descending sides for such systems are essentially enantiographic, and this is true for more complicated (but slowly equilibrating) systems such as $m\mathrm{A}+n\mathrm{B}\rightleftharpoons\mathrm{C}$, where three peaks would be observed in each of the conjugate ascending and descending patterns. Cases do exist, however, where the velocities of all species in slow equilibrium are identical in a particular type of migration process. For example, urease exhibits three peaks (corresponding to monomer, dimer, and trimer) in velocity sedimentation, but only single boundaries in the ascending and descending electrophoretic patterns (Creeth and Nichol 1960). Slowly equilibrating systems may equally well be studied by zonal migration methods (for example, Fig. 2.1(a)), but the opportunity of comparing conjugate sides is then lost.

5.2. Pressure effects and the rectangular approximation in velocity sedimentation

In comparison with other mass migration methods, velocity sedimentation has at least three disadvantages: (i) the lack of the conjugate boundary system, and consequent inability to yield information on enantiography; (ii) the use of sector-shaped cells and inhomogeneous centrifugal fields in contrast with the cells of uniform cross-section and homogeneous fields for which the theory has been developed; and (iii) the development of high pressures (100–500 atm) at the bottom of the cell. There is little that can be done about the first disadvantage, since a conjugate boundary system would be gravitationally unstable without the superimposition of a gradient (sucrose, LiCl, etc.). While considerable success has been achieved with such stabilization in zonal sedimentation of non-interacting systems, the introduction of an additional component (and hence the possibility of additional interactions) seems unwise in investigations of solute-solute interactions.

(a) *The rectangular approximation in velocity sedimentation*

In the preceding chapters the theory for the migration of rapidly interacting systems has been developed for migration in a cell with constant cross-sectional area, subjected to a uniform field. Implicit in its direct application to sedimentation velocity results is the 'rectangular approximation', i.e. the neglect of the sectorially-shaped cell and the inhomogeneous field. Qualitatively it is clear that both of these factors lead to a progressive dilution of solute during migration of the boundary from the meniscus to the cell bottom. As a consequence, the mean boundary position is no longer given by the first moment of the boundary (eqn (3.4)) but by the square root of the second moment (Goldberg 1953, Trautman and Schumaker 1954), i.e.

$$\bar{x}^2 = \frac{\int x^2(\mathrm{d}\bar{c}/\mathrm{d}x)\,\mathrm{d}x}{\int (\mathrm{d}\bar{c}/\mathrm{d}x)\,\mathrm{d}x}, \tag{5.2}$$

the limits of integration being the solvent and original solution plateaux. Furthermore, interpretation of the values of $\bar{s}$ obtained for a polymerizing system from the rate of movement of $\bar{x}$ is complicated by the fact that the plateau concentration is decreasing progressively according to eqn (5.3),

$$\bar{c}_x = \bar{c}_o(x_m/x)^2, \tag{5.3}$$

where $\bar{c}_x$ refers to the plateau concentration ($\bar{c}_o$ initially) after migration of the boundary from the meniscus, x_m, to a position x: an average of the $\bar{c}_x$ values in the time interval required for measurement of $\bar{s}$ is usually taken as $\bar{c}$. By this means a series of ($\bar{s}$, $\bar{c}$) values may be obtained for interpretation according to eqn (3.8), as described in Chapter 3.

It was also noted in Chapter 3 that equilibrium constants for monomer-single higher polymer systems could be obtained from the concentration of the slower peak ($\bar{c}_{min}$) in the bimodal reaction boundary observed when $n > 2$ (eqn (3.7)). Townend et al. (1960) have, for example, reported a value of K calculated from $\bar{c}_{min}$ for β-lactoglobulin A. Consideration of a model which incorporates the effects of radial dilution but still neglects diffusion (Fujita 1962, Payens and Schmidt 1966) leads to the following expression for K (Nichol *et al.* 1964):

$$K=(\bar{c}_{\text{min}})^{1-n}\,\frac{(n-2)\psi\{n(2n-1)+(n-2)\psi\}^{n-1}}{\{n(2n-1)\}^{n}}, \tag{5.4}$$

where
$$\psi=\frac{1-\dfrac{s_{\text{C}}}{3ns_{\text{A}}}(n-2)\tau}{1+\dfrac{(2n-1)\tau}{3}} \tag{5.5}$$

and
$$\tau=2s_{\text{A}}\,\omega^{2}t. \tag{5.6}$$

The limit of ψ as $t\to 0$ is unity, whereupon eqn (5.4) reduces to eqn (3.7), used by Townend *et al.* (1960). However, the subsequent decrease of ψ with increasing t would lead to progressively smaller estimates of K. Application of eqn (5.4) instead of eqn (3.7) would reduce the estimate of the association constant from $3\times10^{11}\ \text{l}^3\ \text{mol}^{-3}$ to $2\times10^{11}\ \text{l}^3\ \text{mol}^{-3}$: these values correspond to standard free energy changes of $-14{\cdot}6$ and $-14{\cdot}4$ kcal/base mol, respectively. Such a change is undoubtedly of little significance from the practical viewpoint, particularly when (a) the difficulties encountered in measuring areas of partially resolved peaks and (b) the continued neglect of diffusion in the revised theoretical treatment are taken into account.

(b) *Effects of pressure in velocity sedimentation*

In ultracentrifugation the pressure, P, varies with the radial distance from the centre of rotation, x, according to the expression $\mathrm{d}P=x\omega^2\rho\ \mathrm{d}x$, where ω is the angular velocity and ρ the density of the solution. Thus, the pressure dependence of the equilibrium constant for (say) a polymerizing system involving a molar volume change on reaction of ΔV, viz. $\mathrm{d}\ln K_n/\mathrm{d}P=-\Delta V/RT$, may be expressed (Josephs and Harrington 1966, Ten Eyck and Kauzmann 1967) in the integrated form as

$$K_n(x_1)=K_n(x_2)\mathrm{e}^{-\{\Delta V\omega^2\rho(x_1{}^2-x_2{}^2)\}/2RT}. \tag{5.7}$$

Howlett, Jeffrey and Nichol (1970, 1972) have discussed evaluation of ΔV from sedimentation equilibrium experiments for which eqn (5.7) applies even though the pressure gradient and hence K variation are generally less pronounced than in sedimentation velocity.

A practical example of the effects of pressure in sedimentation velocity is provided by the studies of Josephs and Harrington (1968) on the association of myosin, for which ΔV is positive. The resultant conversion of myosin polymers to monomeric protein at high pressures leads to unusual effects during velocity sedimentation, as is evident from Fig. 5.2. Frame (a) is the pattern obtained at relatively low speed (24 000 rev/min), where the pressure effect is too small to perturb appreciably the equilibrium, with the result that boundaries corresponding to monomer and polymer are observed. Increase of the speed to 60 000 rev/min generates sufficient pressure for the equilibrium to be displaced completely in favour of monomer at all points throughout the solution in the cell, whereupon the 'polymer' boundary is transformed into a differential boundary between two concentrations of monomer. The subsequent progress of the run is summarized in frames (b)–(f) of Fig. 5.2, from which it is evident that the monomer peak overtakes and merges with the differential boundary: numbers adjacent to peaks indicate the s values of the various boundaries. Further proof that the 3·6 S peak is indeed a differential boundary is provided by the very good agreement between the observed sedimentation coefficient and the value of 3·7 S predicted on the basis of eqn (5.8), the expression describing the migration of such a boundary (Schachman 1959),

$$s_\Delta = \frac{c_2 s_2 - c_1 s_1}{c_2 - c_1}. \tag{5.8}$$

Subscripts 1 and 2 refer to regions to the centripetal and centrifugal sides, respectively, of the differential boundary.

Although the data shown in Fig. (5.2) may be considered to be dramatic consequences of pressure effects, simulated data (Kegeles, Rhodes, and Bethune 1967) for interacting systems suggest that even more bizarre patterns might arise. By an adaptation of the computer programme for a countercurrent distribution analogue (Bethune and Kegeles 1961*c*), modified to incorporate the effects of pressure on K, it has been shown, for example, that as many as five peaks (four positive, one negative) may be predicted for a system of the type $A + B \rightleftharpoons C$. In this connection, the existence of negative concentration gradients occurred frequently in the simulated examples, particularly those for polymerizing systems in which pressure favoured dissociation, where convective disturbances were invariably predicted. Although the experimental patterns of Josephs and Harrington (1966, 1967, 1968) seem to be free from such effects, Kegeles (1969) has shown that convection must be a general phenomenon during velocity sedimentation of associating systems for which the pressure-induced perturbation of the equilibrium leads to the formation of monomer. The argument is as follows.

Consider the ultracentrifuge patterns shown for myosin in frames (b) and (c) of Fig. 5.2. As stated earlier, the solution plateau on the centrifugal side of the differential boundary corresponds to monomeric protein, while the peak centripetal to it represents a boundary between solvent and monomer. Since the plateau solution comprises monomeric myosin, it follows that the position of the equivalent sharp boundary obtained from second moment calculations (eqn (5.2)) across the entire pattern (solvent to solution) must coincide with the centre of the monomer boundary (5·4 S peak): this can only occur if there are balancing positive and negative concentration gradients between the

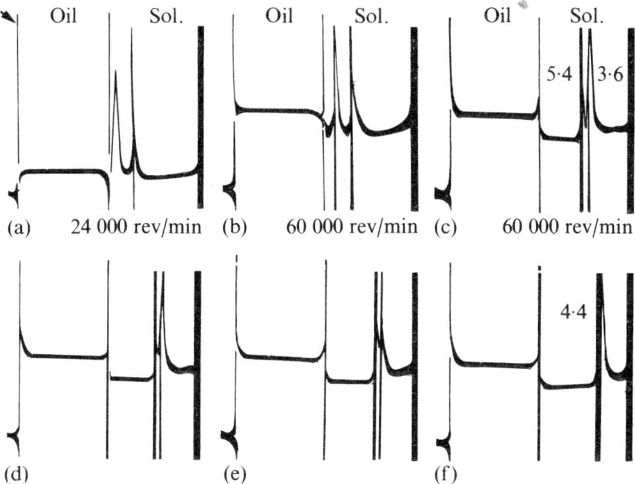

FIG. 5.2. Effect of pressure in velocity sedimentation of myosin, the polymerization of which involves a molar volume increase. Frame (a), taken at low speed (24 000 rev/min), reflects *essentially* the situation pertaining under atmospheric conditions, while frames (b)–(f) demonstrate the breakdown of polymeric species under the pressures operating at a rotor speed of 60 000 rev/min: the thickness of the oil layer was selected to increase the pressure sufficiently for complete conversion to monomer. (Adapted from Josephs and Harrington (1968).)

monomer boundary and the original solution plateau region. Such a system would be gravitationally unstable, the resulting convection giving rise to the observed pattern in which the density increases continuously with increasing distance from the centre of rotation.

To conclude this section, it is clear that the use of velocity sedimentation should be avoided, if possible, in the study of interactions *which are accompanied by appreciable volume changes.* Even in the event that it proves possible to allow for the effect of the pressure-dependence of the equilibrium constant on the sedimentation behaviour, the possibility of predicting accurately the consequences of superimposed convective disturbances seems very remote.

This factor has prompted the suggestion by Kegeles *et al.* (1967) that velocity sedimentation velocity results be no longer taken as acceptable evidence for any conclusions on interacting systems, unless supported by auxiliary data that are free from effects of pressure. On the other hand, sedimentation velocity results obtained with several self-interacting globular protein systems (for example, insulin, β-lactoglobulin, α-chymotrypsin, lysozyme, sulphatase A, and haemoglobin) have been supported by other methods devoid of pressure effects.

5.3. Effects of charge

In any mass migration experiment, be it of the freely migrating or chromatographic variety, the existence of a net charge on a macromolecule must modify to some extent its velocity compared with that of an uncharged but otherwise identical solute. Clearly these effects should be most pronounced in electrophoresis, where the solute migrates under the influence of an electric field.

(a) *Dependence of electrophoretic mobility on charge and ionic strength*

The relationship between electrophoretic mobility, U (cm^2 s^{-1} V^{-1}) and net charge (or valence) z has not been established unequivocally, but eqns (5.9a) (Henry 1931) and (5.9b) (Abramson *et al.* 1942) have been postulated for spherical particles.

$$U = \left\{\frac{f(\kappa a)}{6\pi\eta a(1+\kappa a)}\right\} z \times 1{\cdot}60 \times 10^{-10}, \tag{5.9a}$$

$$U = \left\{\frac{f(\kappa a)(1+\kappa a_i)}{6\pi\eta a(1+\kappa a+\kappa a_i)}\right\} z \times 1{\cdot}60 \times 10^{-10}. \tag{5.9b}$$

η is the viscosity of the medium (centrepoises) and $1/\kappa$ corresponds to the thickness of the Debye–Huckel ion atmosphere: a and a_i denote the radii of the macroion and supporting electrolyte ions, respectively, while $f(\kappa a)$ is the Henry function. At constant ionic strength both of the terms within the braces approximate to constants for proteins free from pH-dependent changes in size and/or shape, and hence linear U–z relationships are predicted by either expression. Electrophoretic mobility (Longsworth 1941) and titration (Cannan, Kibrick, and Palmer 1941) results on ovalbumin illustrate such dependence of mobility on charge at constant ionic strength (Fig. 5.3(a)). However, the agreement between experiment and theory is only qualitative since the experimental slope (solid line) differs markedly from that of the dashed line, which represents the theoretical plot for eqn (5.9b). This discrepancy may reflect either the limitations of the approximations inherent in the theoretical treatments of electrophoretic migration (Booth 1950, Overbeek 1950, Barnett and Bull 1960) and/or to erroneous estimates of z, which were based on pH-titration data uncorrected for salt-ion binding (see, for example, Waldmann-Meyer 1960): in this connection the triangles in Fig. 5.3(a) denote values of z corrected for chloride binding (Carr 1953), while the square (from data of Creeth and Winzor 1962) is based on estimates of charge by the Charlwood (1950) procedure.

In addition to being a function of charge, U also varies with ionic strength (I) of the medium. The magnitude of $f(\kappa a)$ is relatively insensitive to changes of ionic strength, varying from 1·0 at low salt concentrations to 1·5 at high ionic strengths: most of the mobility variation stems from the dependence of κ on $\sqrt{I}$. Figure 5.3(b) summarizes the experimental results of Tiselius and Svensson (1940) for ovalbumin at pH 7·1, together with the theoretical relationships between U and $\sqrt{I}$ on the basis of eqns (5.9a) (solid line) and (5.9b) (broken line). Again, the agreement between experiment and prediction should only be considered as qualitative.

In studies of chemically reacting systems by moving boundary electrophoresis the above effects of charge and of ionic strength on the mobilities of the interacting species will obviously influence greatly the *nature* of the

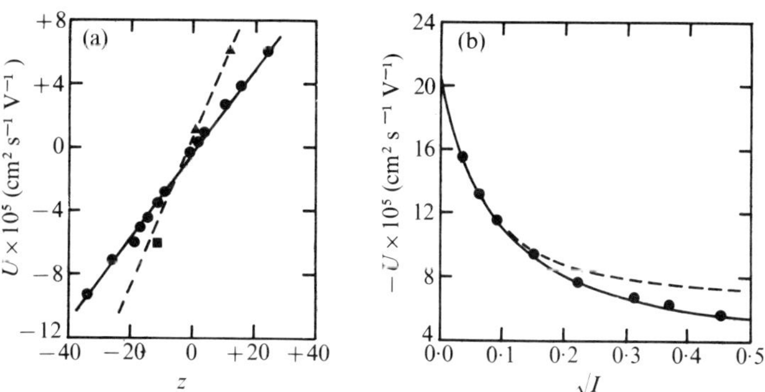

FIG. 5.3. Effects of charge and ionic strength on the electrophoretic mobility of ovalbumin. (a) Dependence of mobility on net charge (I=0·1), the broken line representing the relationship predicted by eqn (5.9b): ●, values inferred from Table 1 of Longsworth (1941), the solid line being drawn through these points; ▲, results of Longsworth corrected for chloride binding; ●, data of Creeth and Winzor (1962). (b) Mobility in phosphate buffer solutions of pH 7·1 and varying ionic strengths: experimental points are taken from Table 1 of Tiselius and Svensson (1940), the solid and broken lines being the theoretical curves predicted by eqns (5.9a) and (5.9b), respectively.

migration patterns. For example, a particular reactant may be the faster migrating solute under one set of conditions (pH) but the slower under another. Caution should therefore be exercised in the interpretation of electrophoretic data on interacting systems, particularly in view of the fact that variations in pH and ionic strength occur in electrophoresis of non-interacting systems (section 5.3(b)).

(b) *The δ and ε boundaries in moving boundary electrophoresis*

According to the general theoretical treatment of the migration of a single non-interacting solute (Chapter 2), the ascending and descending electrophoretic patterns should each comprise a single boundary at the position $x=ut$ (Fig. 2.2). Inspection of experimental results does not bear out this prediction, since additional concentration gradients, the δ and ε boundaries, are observed in the vicinity of the positions occupied by the initial interfaces between solution and solvent. These anomalous boundaries result from the requirement that the electric current at a particular position in the cell is time-independent. Such a requirement is expressed mathematically by constancy of the Kohlrausch (1897) regulating function, Ω, defined as

$$\Omega=\sum\frac{m_i z_i}{U_i} \tag{5.10}$$

in all phases of the electrophoretic system (Longsworth 1947, 1959*a*), represented schematically as

descending side: S, R, A(α)$\leftarrow$S, R(β) : : S, R(γ),

ascending side: S, R(c)$\leftarrow$S, R, A(b) : : S, R, A(a),

where S and R represent cations and anions, respectively, of the supporting electrolyte, A the single unreacting solute, and letters in parentheses phases in the cell: the arrows denote moving boundaries, while the approximately stationary *ab* and $\beta\gamma$ boundaries refer to the δ and ε boundaries, respectively. From eqn (5.10) the solute thus contributes to Ω in the α or *a* plateau region, whereas the magnitude of Ω in the γ or *c* regions is governed solely by the concentrations, etc. of the supporting electrolyte. Provided z_i and U_i are considered to be independent of phase, there must be a redistribution of solute and solvent ions for the regulating functions of all phases to be equated (eqn (5.11)):

$$\frac{m_A^\alpha z_A}{U_A}+\frac{m_S^\alpha z_S}{U_S}+\frac{m_R^\alpha z_R}{U_R}=\frac{m_S^\beta z_S}{U_S}+\frac{m_R^\beta z_R}{U_R}=\frac{m_S^\gamma z_S}{U_S}+\frac{m_R^\gamma z_R}{U_R}. \tag{5.11}$$

Across the anomalous boundaries, which reflect these redistributions of charged species, the concentrations of all ions remain in the same relative proportions, the dilution factor being governed by the ratio $\Omega^\alpha/\Omega^\gamma$. Longsworth (1947, 1959*a*) has verified the validity of these expressions by predicting the magnitude of the ε boundary on the basis of z_A and the Donnan (1923) expression relating the concentrations of electrolyte ions in the α and γ (diffusate) phases. Theoretically these concentration boundaries should remain stationary, their slight movement reflecting the approximation that all U_i are constants.

Because of the existence of the δ and ε boundaries it follows that the descending pattern gives a measure of the mobility of a single solute under the conditions prevailing in the α phase, while its ascending counterpart records mobility in a more dilute environment. From Fig. 5.3(b) we should thus expect non-enantiography of electrophoretic patterns even for systems devoid of chemical interactions. An additional source of mobility differences between conjugate limbs stems from the fact that most electrophoresis is performed in buffer systems containing weak acids. Evidence (Longsworth 1947, 1959*a*) suggests that the concentration of undissociated acid HR remains invariant during electrophoresis, whereupon the dilution of buffer ions that occurs in the *b* plateau region must also induce a change in pH: this effect accounts at least partly for the relatively greater size of the δ boundary compared with that of its descending counterpart.

The extent to which the ascending and descending patterns differ experimentally for a non-interacting solute system may be gauged from Fig. 5.4(a), (b), which illustrate results obtained with ovalbumin at different ionic strengths (Cann 1949): although the effect is more pronounced at the lower ionic strength (0·05), it is still significant experimentally at an ionic strength of 0·1. Not only are the ascending and descending patterns non-enantiographic because of the faster migration rate of the former, but also because of the relative sharpness and hence better resolution observed in the ascending boundary. This difference in shape reflects the changes of conductivity and pH that occur across the $\alpha\beta$ and *bc* boundaries. If the mobility of a macromolecular cation is less than that of the electrolyte cation, the change in conductivity is such that solute molecules in the leading edge of the *bc* boundary are migrating more slowly than those in the *b* region, with a consequent sharpening of the *bc* boundary: the conductivity gradient in the $\alpha\beta$ region leads to a corresponding increase in the extent of boundary spreading in the descending pattern (Svensson 1943). Variations in pH can either enhance or counter to some extent the effects of changes in conductivity (Longsworth 1959*a*).

Despite the degree of non-enantiography introduced by the above effects it must be stressed that this is frequently of minor magnitude compared with that consequent upon chemical interaction. For example, Fig. 5.4(c) shows the hypersharp ascending and spread bimodal descending boundaries obtained in the electrophoresis of β-lactoglobulin: there is no doubt that these patterns reflect mainly the self-association of the protein.

From this discussion of the stationary boundaries it is clear that although moving boundary electrophoresis has the advantage of yielding ascending as well as descending boundary systems, quantitative interpretation is hindered by the difficulty of defining the operative environment for all boundaries except that which migrates fastest in the descending limb. The consequent errors in estimations of concentrations and mobilities has been ascertained from experimental (Cann 1949) and simulated (Longsworth 1959*a*) results on non-interacting mixtures. In electrophoretic investigations of interacting systems (for example, Longsworth and MacInnes 1942, Cann and Klapper 1961) the absence of ion redistribution must be assumed, and consequently results from the descending side are preferable from the viewpoint that conditions approximate more closely to this requirement (Longsworth

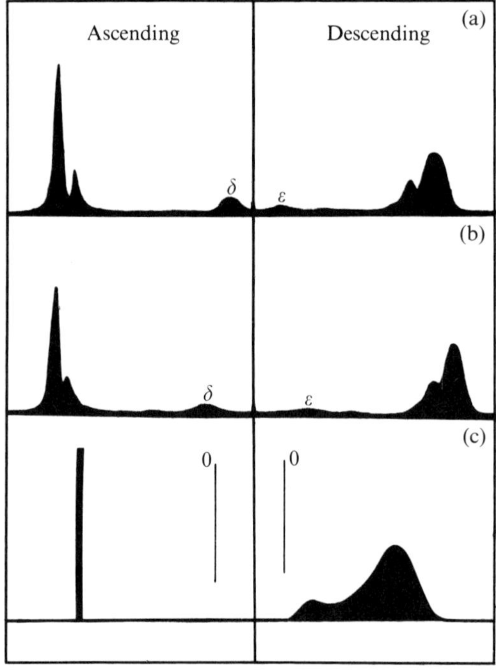

FIG. 5.4. Non-enantiography of ascending and descending profiles in moving boundary electrophoresis: (a) ovalbumin in 0·05 I phosphate, pH 6·8; (b) ovalbumin in 0·1 I phosphate, pH 6·8; (c) salt-fractionated β-lactoglobulin in 0·1 I acetate, pH 4·66 (original boundary positions denoted by O). ((a) and (b) adapted from Cann (1949), (c) from Tombs (1957).)

1959*a*). With this approximation the data may then be interpreted in terms of the composition of the solution subjected to electrophoresis.

(c) *Charge and salt effects in velocity sedimentation*

More indirect, and frequently less pronounced, effects result from the electrostatic interactions that occur whenever a macroion and the ions of the supporting electrolyte migrate at different speeds. A potential gradient is thereby established which may either increase or decrease the migration rate of the macromolecule according to the direction of this additional driving force in relation to that of the ultracentrifugal field. In the case of the primary charge effect, which is observed at very low ionic strengths, the electrostatic field results from the differential sedimentation of an ionized macromolecule and its counterions, and is therefore in opposition to the centrifugal force. A secondary effect stems from differences between the migration rates of the anions and cations of the supporting electrolyte: depending on the charge of the macroion, the resulting potential gradient may either enhance or oppose its sedimentation. Our understanding of these primary and secondary charge effects is due largely to the theoretical studies of Pedersen (Svedberg and Pedersen 1940, Pedersen 1958), and Booth (1954), the former having also performed the experiments required to demonstrate the existence and importance of these phenomena in velocity sedimentation. Although there have been no reports of similar effects in gel chromatography, there seems to be no reason for doubting their existence.

Consider the centrifugation of a solution containing an ionized macromolecular solute and a number of other ions with valences z_i, concentrations m_i, sedimentation coefficients s_i, and electrophoretic mobilities U_i: the subscript 'i' = A is used to denote the macroion, and 'i' = 1, 2, . . . *n* for the other ions. Because of the different magnitudes of *s* for the various species, an electrical and a centrifugal field pertain, giving rise to velocities $v_A, v_1, v_2, \ldots v_n$. In accordance with the theory suggested by Tiselius (1932) these velocities are considered to represent the algebraic sum of sedimentation and electrophoretic terms (eqn (5.12)).

$$\begin{aligned}
v_A &= s_A\,\omega^2 x - U_A(\mathrm{d}\psi/\mathrm{d}x),\\
v_1 &= s_1\,\omega^2 x - U_1(\mathrm{d}\psi/\mathrm{d}x),\\
&\;\,\cdot\\
&\;\,\cdot\\
&\;\,\cdot\\
v_n &= s_n\,\omega^2 x - U_n(\mathrm{d}\psi/\mathrm{d}x),
\end{aligned}\tag{5.12}$$

$U_A, U_1, \ldots U_n$ are the electrophoretic mobilities of the ions, while $s_A, s_1, \ldots s_n$ refer to the sedimentation coefficients of equivalent uncharged particles: $d\psi/dx$ denotes the potential gradient that is established. These expressions may be combined with the condition of electroneutrality ($\sum m_i z_i v_i = 0$) to yield the following relationship:

$$\frac{d\psi}{dx} = \frac{\omega^2 x \sum m_i z_i s_i}{\sum m_i z_i U_i}. \tag{5.13}$$

Thus the observed sedimentation coefficient of the macroion, $s'_A = v_A/\omega^2 x$, is given by

$$s'_A = s_A - U_A \frac{\sum m_i z_i s_i}{\sum m_i z_i U_i}. \tag{5.14}$$

The *primary charge effect* is observed at very low ionic strengths, such that $\sum m_i z_i s_i \simeq m_A z_A s_A$; whereupon eqn (5.14) may be rearranged to give

$$s'_A = s_A \left\{ 1 - \frac{m_A z_A U_A}{\sum m_i z_i U_i} \right\}. \tag{5.15}$$

Since z_i and U_i always carry the same sign (by convention) it follows that the term within parentheses is always positive, and hence the observed sedimentation coefficient s'_A is always less than s_A, but approaches that value as the mobility term tends to zero. Thus there are three procedures available for reducing a primary charge effect to negligible proportions, viz. (a) reduction of the macroion concentration m_A, (b) selection of a pH in the region of the isoelectric point (U_A and $z_A \rightarrow 0$), and (c) addition of a suitable neutral salt to increase the $\sum m_i z_i U_i$ term. Its complete suppression by the third means, termed the *primary salt effect*, can only be accomplished if the summation $\sum m_i z_i s_i$ for the added electrolyte is zero. Otherwise the quantity $(s'_A - s_A)$ may be either positive or negative, depending on the direction of the electric field created and the sign of the charge on the macromolecule. Clearly, this *secondary salt effect* is enhanced, not suppressed, by increasing the concentration of added electrolyte.

Examples showing the existence of these three phenomena are given in Fig. 5.5, which summarizes results obtained by Pedersen (1958) for bovine serum albumin. Figure 5.5(a) illustrates the sedimentation behaviour of electrodialysed serum albumin solutions to which have been added increasing amounts of HCl. The initial increase in s'_A is expected, since the electrodialysed solution bore a net negative charge, whereupon the initial addition of acid resulted in a reduction rather than an increase of charge. Although these results are thus qualitatively in accord with theory, the value of s'_A decreases far below that predicted by eqn (5.15), presumably because of the structural changes undergone by serum albumin in acidic solution (Charlwood and Ens 1957). Suppression of the primary charge effect by the addition of a neutral salt (LiBr) is shown in Fig. 5.5(b), from which it is evident that an increase of the ionic strength from 0·05 to 0·20 produces very little change in the value of s'_A (suitably corrected for density, viscosity, and temperature): thus the primary charge effect is virtually eliminated by addition of 0·05 M salt in this case. However, different limiting values of s'_A are obtained with LiBr, KBr, and CsBr

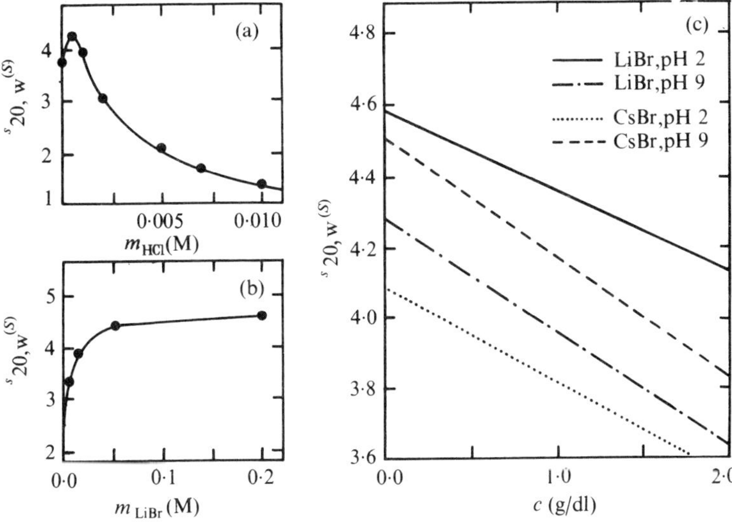

FIG. 5.5. Charge and salt effects in velocity sedimentation of bovine serum albumin. (a) Sedimentation coefficients of 1 per cent albumin solutions in varying concentrations of HCl (primary charge effect); (b) effect of added electrolyte (LiBr) on $s_{20,w}$ for 1 per cent albumin solutions pH 2 (primary salt effect): (c) dependence of the s–c relationship for serum albumin upon the nature of the added electrolyte (secondary salt effect). (Adapted from Pedersen (1958).)

as the supporting electrolyte, this being a manifestation of the secondary salt effect. With LiBr as added salt the superimposed potential gradient should speed up the sedimentation of cationic macromolecules, whereas CsBr should favour the sedimentation of anionic species. Inspection of Fig. 5.5(c) confirms this prediction: in acid solution the sedimentation coefficient of bovine serum albumin is clearly greater in LiBr than in CsBr, the opposite effect being observed at alkaline pH, where the protein bears a net negative charge.

From the viewpoint of investigating chemically reacting systems the fact that some or all of the reactant and product velocities may differ from those of uncharged counterparts would not usually influence unduly the results obtained, unless one of the reacting species were contributing substantially to the ionic strength of the system. Under those circumstances allowance would have to be made for the variations in velocities of species in different regions of the boundary system. An environmental factor of this type may be construed as a physical interaction, the treatment of which is considered in section 5 of this Chapter.

(d) *Gibbs–Donnan effects*

Inasmuch as many mass-migration experiments on interacting systems are preceded by a dialysis step, some mention should be made of the consequences of equating the chemical potentials of macroion and buffer solutions by dialysis equilibrium. Such phenomena are obviously operative regardless of the method by which the dialysed solutions are studied; but in gel chromatography there is an additional Gibbs–Donnan effect resulting from the exclusion of macromolecular solutes from regions of the stationary phase (Nichol, Sawyer, and Winzor 1969*a*). The effects of dialysis equilibrium will be considered first.

(i) *Consequences of dialysis equilibrium.* Consider an aqueous solution of a macroion A bearing z_A negative charges, which are counterbalanced by cations S to preserve electroneutrality. After exhaustive dialysis of the solution against an aqueous solution of uniunivalent electrolyte SR the Gibbs–Donnan equilibrium for the system may be expressed (Longsworth 1959*a*):

$$(m_S)_i = (m_S)_o + \tfrac{1}{2} z_A m_A, \tag{5.16a}$$

$$(m_R)_i = (m_R)_o - \tfrac{1}{2} z_A m_A, \tag{5.16b}$$

m_A is the molar concentration of macroion, m_S and m_R being the total molar concentrations of S and R, respectively: the subscript 'i' distinguishes concentrations within the dialysis sac from those in the diffusate (subscript 'o'). From eqn (5.16b) it follows that $(m_R)_i < (m_R)_o$, and thus the concentration of unbound SR within the dialysis sac is less than its concentration in the diffusate (eqn (5.17)):

$$(m_{SR})_i = (m_{SR})_o - \tfrac{1}{2} z_A m_A. \tag{5.17}$$

Irrespective of whether the electrolyte concentration within the sac is considered to be $(m_S)_i + (m_R)_i$ or $(m_{SR})_i$, its concentration in the diffusate does not represent that in the solution unless the macromolecule bears zero charge. Since the extent of this difference in ionic strength depends on m_A, exhaustive dialysis of different concentrations of macroion against the same buffer does not ensure identity of conditions within the various dialysis sacs. The consequent uncertainty as to the exact conditions within the dialysis sac is not confined to the ionic strength, since any undissociated acid HR (present to provide buffering capacity) distributes uniformly across the dialysis membrane: minor pH differences between the various solutions of dialysed macroion and the diffusate could result. Thus while dialysis guarantees the existence of reproducible conditions within a solution of a macroion, it also ensures uncertainty as to the exact conditions that pertain in the dialysed solution. Clearly, knowledge of the concentrations and valences of all macromolecular reactants and products in the solution is required for application of eqn (5.17) to define the environment precisely.

As well as establishing the existence of this element of doubt produced by exhaustive dialysis, eqn (5.17) also provides the information required for reducing to negligible proportions the magnitude of this uncertainty. Since the absolute increment in the concentration of diffusible ion across the membrane is independent of its concentration, an increase in the concentration of supporting electrolyte leads to a corresponding decrease in the relative concentration difference. For example, consider dialysis of a 1 per cent solution of protein, molecular weight 50 000 and valence 10. The magnitude of the difference in electrolyte concentration across the membrane is 0·002 M: in exhaustive dialysis against 0·01 M, 0·1 M, and 1·0 M NaCl solutions the relative difference between the concentrations of salt in the solution and in the diffusate would be 20, 2, and 0·2 per cent, respectively. Thus, provided the chemical equilibrium under investigation is not a weak electrostatic interaction observable only at very low ionic strength, then the Gibbs–Donnan effects resulting from predialysis of macromolecular solutions may be made to assume negligible proportions by suitable choice of ionic strength. In the event that a weak electrostatic interaction is the reaction of interest, a preferable procedure for the preparation of reaction mixtures would be the addition of electrolyte and other reactants to electrodialysed solutions of macroions. But having done so the experimenter would still face the prospect of dealing with charge effects in electrophoresis (section 5.3(a), velocity sedimentation (section 5.3(b)) or gel chromatography (see below), any of which could be substantial under conditions of low ionic strength.

(ii) *The Gibbs–Donnan equilibrium in gel chromatography.* The various models of gel chromatography may be divided into two broad categories, viz. those based on paritition equilibrium of solute between the mobile phase and stationary liquid phase within the gel matrix (Porath 1963, Squire 1964, Laurent and Killander 1964), and those based on non-equilibrium between phases (Ackers 1964, Giddings and Mallik 1966). Fortunately, gel chromatographic profiles exhibit relative insensitivity to column flow rate (Flodin 1962;

Ackers 1964; Siegel and Monty 1966; Hibberd, Ogston, and Winzor 1970), and thus vindicate the use of simpler equilibrium models as adequate descriptions of the gel chromatographic process. Common to all of the equilibrium concepts of gel chromatography is the supposition that solute is uniformly distributed between the mobile phase and *accessible* regions within the gel, the magnitude of the distribution coefficient reflecting the extent to which a solute penetrates the stationary phase. Existence within the gel of regions devoid of large but not smaller ions renders the system analogous to the situation operative across a semipermeable membrane, and thus osmotic and Gibbs–Donnan effects must be expected to occur: both have, indeed, been observed with Sephadex gels (Edmond, Farquhar, Dunstone, and Ogston 1968; Nichol *et al.* 1969*a*).

Redistribution of the supporting electrolyte (and any small reactant ion) is described by the Gibbs–Donnan expression (eqn (5.17)), with $(m_{SR})_i$ now referring to the concentration of small solute in the mobile phase (and in the portion of gel phase accessible to larger solute), and $(m_{SR})_o$ to its concentration in the region free from macromolecular solute. In the event that the larger ion should be distributed throughout the entire stationary phase but located preferentially in the mobile phase (Ackers 1964), the Gibbs–Donnan expression (eqn (5.17)) would still apply, but with m_A replaced by Δm_A, the difference between concentrations of macroion in the two phases. In order to investigate the effects of Gibbs–Donnan equilibria in gel chromatography, a model system was chosen (Nichol *et al.* 1969*a*) in which the macroion (dextran sulphate) was completely excluded from the gel (Sephadex G-25) phase, thereby eliminating any confusion as to the significance of the parameters $(m_{SR})_i$ and $(m_{SR})_o$.

Consider the situation pertaining in frontal gel chromatography of a dialysed solution of macromolecular anion A in electrolyte SR. Provided it may be assumed that S and R have the same elution volume (V_{SR}), then use of water to elute a column equilibrated previously with the dialysed solution of macroion gives rise to the following boundary distribution:

$$AS_n,\ S,\ R,\ H_2O(\alpha) \leftarrow S,\ R,\ H_2O(\beta) \leftarrow H_2O(\gamma).$$

We now write a moving boundary equation to describe the mass conservation of a constituent that does not disappear across the $\alpha\beta$ boundary, choosing R so that the expression then describes the distribution of unbound SR. The appropriate equation, referring to the elution profile (denoted by subscript 'e'), is (Nichol *et al.* 1967*b*)

$$\left(V_o - \frac{V_o}{\mu_{SR}^{\alpha}}\right)(m_{SR}^{\alpha})_e = \left(V_o - \frac{V_o}{\mu_{SR}^{\beta}}\right)(m_{SR}^{\beta})_e, \tag{5.18}$$

where μ_{SR}^{α} represents the ratio of the amount of free SR in the mobile phase to the total free SR in both mobile and stationary phases of the α plateau region on the column: V_o, the void volume of the column, is also the elution volume of the macroion, since μ_A^{α} is unity. If subscripts 'm' and 's' are used to denote concentrations in the mobile and stationary phases, respectively, it follows that

$$\mu_{SR}^{\alpha}=\frac{V_o(m_{SR}^{\alpha})_m}{\{V_o(m_{SR}^{\alpha})_m+(V_{AR}-V_o)(m_{SR}^{\alpha})_s\}}, \tag{5.19}$$

whereas

$$\mu_{SR}^{\beta}=\frac{V_o}{V_{SR}} \tag{5.20}$$

since the concentrations of SR in the mobile and stationary phases are identical in the absence of macroion. Combination of these three expressions yields

$$(m_{SR}^{\beta})_e=\frac{(m_{SR}^{\alpha})_e(m_{SR}^{\alpha})_s}{(m_{SR}^{\alpha})_m}=(m_{SR}^{\alpha})_s \tag{5.21}$$

noting (section 2.4) that in frontal chromatography of a dialysed solution

$$(m_{SR})_i=(m_{SR}^{\alpha})_m=(m_{SR}^{\alpha})_e. \tag{5.22}$$

In the event that the Donnan equilibrium operates we have already noted that $(m_{SR})_o$ of eqn (5.17) refers to the concentration of SR in the stationary phase, and so the β region of the elution profile records the concentration of small solute in the diffusate, not the dialysed solution. The reality of this situation is evident from Fig. 5.6(a), which depicts the elution profile obtained from a column of Sephadex G-25 equilibrated with a dialysed solution of dextran sulphate: direct dissolution of the dextran sulphate in 0·1 M NaCl yielded the profile shown in Fig. 5.6(b).

Although Fig. 5.6 serves to illustrate the existence of a Gibbs–Donnan effect in gel chromatography, it is emphasized that the error in assuming the identity of $(m_{SR}^{\beta})_e$ and $(m_{SR}^{\alpha})_m$ would be very large in this particular experiment because of the extremely high valence of the dextran sulphate (two charges/glucose residue). Estimations of binding results by equilibrium dialysis and by frontal gel chromatography under conditions where complexed and uncombined macroion move together are thus entirely analogous not only inasmuch as both yield directly the equilibrium concentration of unionized ligand, but also inasmuch as results obtained with charged ligands will be subject to the Gibbs–Donnan effect: the magnitude of this error has already been discussed.

(e) *Neglect of activity coefficients*

Throughout this review of interacting systems the mass action law has been expressed in terms of concentrations instead of activities of the interacting species: implicit therein is the assumption that either the activity coefficient of each reactant and product is unity, or the relevant ratio of activity coefficients is unity. From studies of single, non-associating solutes there is a wealth of evidence to invalidate the former of these alternatives.

Let us first consider a polymerizing system, for which the empirical assumption has usually been made (for example, Adams and Fujita 1963) that the activity coefficient-virial coefficient relationship is given by the expression

$$\ln y_i = \mathrm{i}BM_A\bar{c} \tag{5.23}$$

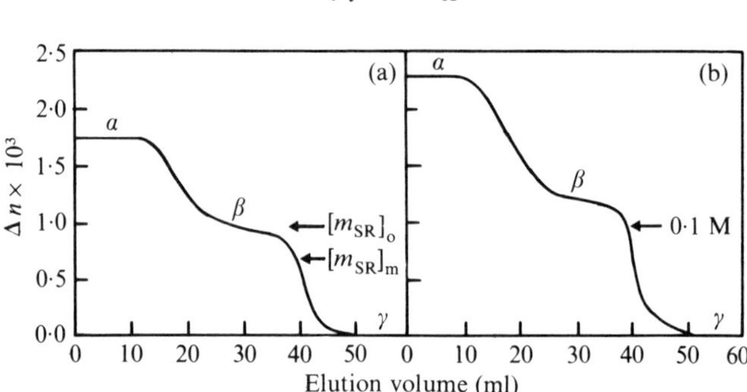

FIG. 5.6. Profiles resulting from elution with water of a 2·5 × 8 cm column of Sephadex G-25 equilibrated with (a) a 1 per cent solution of dextran sulphate dialysed to equilibrium against 0·1 M NaCl, and (b) a 1 per cent solution of dextran sulphate dissolved directly in 0·1 M NaCl (concentrations measured refractometrically). (Adapted from Nichol *et al.* (1969*a*).)

where B is the second virial coefficient common to all oligomer species, and y_i is the activity coefficient of the *i*th species referred to a weight concentration scale. Under these circumstances it is clear that $y_C = y_A^n$ for a system $n\mathrm{A} \rightleftharpoons \mathrm{C}$, whereupon

$$K = \frac{a_C}{a_A^n} = \frac{y_C c_C}{(y_A c_A)^n} = \frac{c_C}{c_A^n}. \tag{5.24}$$

Thus determination of activity coefficients is not essential for the evaluation of association constants describing equilibria between polymers whose activity coefficients conform with eqn (5.23), a formulation which, incidentally, allows their estimation from weight- and number-average molecular weight data.

Recently, Braswell (1968) has provided some justification for this empirical procedure by pointing out that such dependence is likely on the basis of the

Debye–Huckel (1923) theory: in its limiting form the relationship between mean ionic activity coefficients ($y_{\pm}$) and charge may be expressed as

$$\ln y_{\pm} = -\mathbf{A}|z^{+}z^{-}|\sqrt{I}, \tag{5.25}$$

where $|z^{+}z^{-}|$ is the absolute value of the product of the net charges possessed by macroion and gegenion, while **A** is a constant and I the ionic strength. Thus if we consider that no charges are lost in the formation of C from **A**, it follows that

$$(\ln y_{\pm})_{\mathrm{C}} = n(\ln y_{\pm})_{\mathrm{A}}; \qquad \frac{(y_{\pm})_{\mathrm{C}}}{(y_{\pm})_{\mathrm{A}}^{n}} = 1. \tag{5.26}$$

A similar ratio written in terms of activity coefficients expressed on a weight-concentration scale must also equal unity, whereupon eqn (5.23) is justified in these terms. Unfortunately, the situation is less satisfactory for heterogeneous association equilibria ($m\mathrm{A} + n\mathrm{B} \rightleftharpoons \mathrm{C}$) inasmuch as main and cross-linked non-ideality coefficients may have to be considered in expanding $\ln y_{\mathrm{A}}$ and $\ln y_{\mathrm{B}}$ terms (Adams, Pekar, Soucek, Tang, Barlow, and Armstrong 1969): in general, the activity coefficient ratio, $y_{\mathrm{C}}/y_{\mathrm{A}}^{m}y_{\mathrm{B}}^{n}$, may not equal unity except in the limit of infinite dilution (Nichol and Ogston 1965*c*, Adams 1969, Adams *et al.* 1969).

5.4. Diffusion

In Chapter 2 the derivation of the continuity equation, which has formed the basis of succeeding descriptions of migration patterns, neglected terms expressing diffusional flows. The asymptotic profiles thus calculated bear little resemblance to the corresponding experimental data, a point well illustrated by comparing experimental (Rao and Kegeles 1958) and theoretical (Bethune and Kegeles 1961*c*) sedimentation velocity patterns for α-chymotrypsin under conditions where monomer, dimer, and trimer are believed to coexist in rapid association equilibrium (Fig. 5.7(a), (b)). In Figure 5.7(a) is shown the

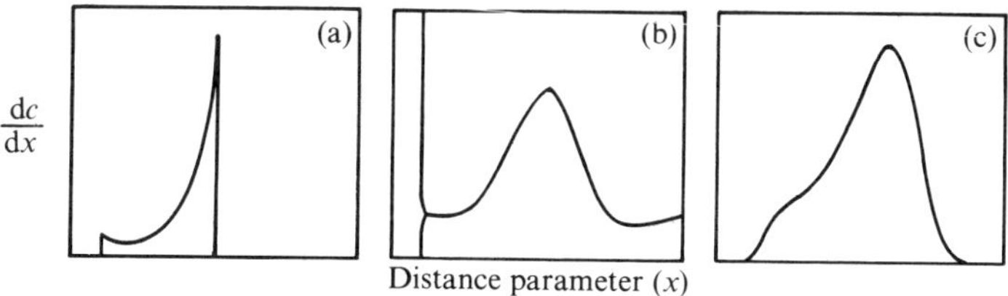

FIG. 5.7. Comparison of theoretical and experimental sedimentation velocity patterns for α-chymotrypsin in 0·2 I phosphate, pH 6·2: (a) the asymptotic profile; (b) the experimental pattern; (c) pattern obtained with the countercurrent distribution analogue of free migration. ((a) and (c) adapted from Bethune and Kegeles (1961*c*), (b) from Rao and Kegeles (1958).)

asymptotic profile calculated on the basis of equilibrium constants derived from molecular weight data (Rao and Kegeles), the experimental result being shown in Fig. 5.7(b). At first sight the two patterns are strikingly dissimilar, as the experimental record displays no sign of the predicted bimodality, an observation which seems to cast doubt on the correctness of the postulated monomer $\rightleftharpoons$ dimer $\rightleftharpoons$ trimer model. However, the slow boundary in the asymptotic profile *is* small, and the investigator is thus faced with the dilemma of deciding whether diffusion could be expected to smear out this boundary to the point of non-detection experimentally. In this connection it is noted that one of the advantages of mass-migration experiments on interacting systems is that they provide not only information on the weight-average velocity of a system, but also a measure of the velocity distribution: as yet this latter potential has not been realized fully in studies of interacting systems because of difficulties encountered in incorporating the effects of diffusional spreading.

Inclusion of diffusional terms in the continuity equations presents no difficulty, but analytical solutions of the resulting expressions can seldom be obtained for systems that exhibit composition-dependent migration. However, provided the phenomenological coefficient for the second-order differential term (D or its analogue) is assumed to be constant, it has been possible to obtain approximate solutions for single solute systems in which the concentration dependence is linear (Hoch 1950, Fujita 1956, Houghton 1963). For more complex concentration dependence of migration alternative procedures are required.

(a) *The countercurrent analogue of free migration*

Bethune and Kegeles (1961*a–c*) attempted to bypass this difficulty by employing a countercurrent distribution analogue to simulate boundary spreading in velocity sedimentation and moving boundary electrophoresis. The theory of countercurrent distribution predicts boundary spreading as a consequence of a series of partitioning steps, and thus there is no need to incorporate a diffusion term into this analogue. In order to simulate a descending boundary system the situation is envisaged wherein the same amount of solute is placed in each tube of the countercurrent train. Coinciding with each quantitative transfer of upper phase a fresh aliquot of upper phase is introduced into the first tube. The corresponding ascending boundary system may also be generated by setting up the countercurrent train with no solute initially, and then adding at each transfer of upper phase an aliquot of upper phase containing a fixed amount of solute to the first tube of the train. Partition and chemical equilibria are attained between transfers, the number of which is selected to give a predetermined degree of resolution. A result of this approach is shown in Fig. 5.7(c), which again refers to the α-chymotrypsin system for which monomer–dimer–trimer equilibria have been postulated. Clearly Fig. 5.7(c) resembles more closely the experimental pattern (Fig. 5.7(b)) than does Fig. 5.7(a), the asymptotic profile. However, the comparison is not entirely satisfactory, possibly due to a pressure effect (Kegeles and Johnson 1970).

The major disadvantage of this method of simulating boundary spreading in freely migrating systems is that the velocity and diffusion parameters cannot be varied independently. Comparison of the equations describing concentration distributions resulting from countercurrent distribution and from free migration shows that the partition coefficient μ (see section 2.4) and $\frac{1}{2}\mu(1-\mu)$ are the analogues of v and D, respectively, N, the number of transfers being the equivalent of t. As a consequence, solutes with the largest velocities must be accorded the largest diffusion coefficients, a situation at variance with usual behaviour: from this viewpoint it is thus preferable to employ numerical solutions of a continuity equation containing the diffusion term.

(b) *Numerical solution of continuity equations containing diffusion terms*

A preferred procedure for making allowance for the effects of diffusional spread is to solve the relevant continuity equation. While analytical solutions have been obtained for very few interacting systems (Hoch 1950, Fujita 1956, Van Holde 1962, Houghton 1963), there is a growing number of studies in which these equations have been solved numerically. In section 1 of this chapter mention was made of the contributions of Cann *et al.* (1957), Cann and Bailey (1961), Scholten (1961), and Belford and Belford (1962), which were all concerned with numerical solution of the continuity equation describing kinetically controlled systems. To these we now add the procedures adopted by Cann and Goad (1965*a*); Dishon, Weiss, and Yphantis (1966, 1967); and Cox (1967, 1969) for the simulation of patterns for solutes undergoing essentially instantaneous interactions. Cann and Goad (1965*a*) were concerned primarily with predicting the electrophoretic behaviour of proteins which reversibly bind undissociated buffer acid, while Dishon *et al.* (1966, 1967) simulated the effects of physical interaction on the sedimentation velocity patterns for single solutes. In addition to treating this latter problem, Cox (1967, 1969) has also concentrated upon the simulation of the sedimentation behaviour of reversibly polymerizing solutes.

In each of these procedures the cell is divided into a series of compartments along the direction of migration, this grid being fixed relative to the x-axis in the Cox (1967, 1969) model, whereas Cann and Goad (1965*a*) have preferred a frame of reference moving with the mean of the velocities of the individual species present: by so doing they were thus able to keep the boundary region(s) located approximately in the middle of the grid. The spacing used by Dishon *et al.* (1967) is even more ingenious inasmuch as the width of compartments varies with the concentration gradient, the value of Δx being decreased in regions of large dc/dx to yield finer detail in the vicinity of boundary regions.

Two different procedures have been adopted for solving numerically the continuity equation pertaining to a system of the type $n\mathrm{A} \rightleftharpoons \mathrm{C}$. Cann and Goad (1965*a*) express the equation of mass balance in the form

$$\frac{\partial(m_\mathrm{A}+nm_\mathrm{C})}{\partial t}=D_\mathrm{A}\frac{\partial^2 m_\mathrm{A}}{\partial x^2}-v_\mathrm{A}\frac{\partial m_\mathrm{A}}{\partial x}+nD_\mathrm{C}\frac{\partial^2 m_\mathrm{C}}{\partial x^2}-nv_\mathrm{C}\frac{\partial m_\mathrm{C}}{\partial x} \tag{5.27}$$

and use the simplest finite difference approximation to this expression for which solution of the finite difference equations approach the solution of the differential equations as Δx and Δt tend to zero. Cox (1967, 1969), on the other hand, has employed average diffusion and sedimentation coefficients ($\bar{D}$ and $\bar{s}$, respectively), defined by eqn (5.28),

$$\bar{D}=\frac{\sum_\mathrm{i} D_\mathrm{i}(\mathrm{d}c_\mathrm{i}/\mathrm{d}x)}{\sum_\mathrm{i}(\mathrm{d}c_\mathrm{i}/\mathrm{d}x)}, \tag{5.28a}$$

$$\bar{s}=\frac{\sum_\mathrm{i} s_\mathrm{i}c_\mathrm{i}}{\sum_\mathrm{i} c_\mathrm{i}}, \tag{5.28b}$$

and has simulated the actual migration by alternating the processes of sedimentation and diffusion. Fig. 5.8 summarizes the results of such computations for a monomer–hexamer system with various values of the initial solute concentration: the predicted constancy of the back peak area (eqn (3.7)) is clearly apparent.

In considering possible consequences of diffusion on conclusions drawn from considerations of diffusion-free migration, Fig. 5.8 serves to illustrate that the major effect is, indeed, a general smearing out of the pattern. Inferences based on the sizes of boundaries are thus not likely to be affected, although quantitative measurement, and possibly detection, of a very small, partially resolved boundary may be virtually impossible with diffusion also taken into account. The weight-average position of a boundary is unaffected by diffusional spreading, and thus measurements of $\bar{v}$, the weight–average velocity, are valid for experimental systems. However, the points of termination of a spread reaction boundary would not be those predicted on the basis of diffusion-free migration ($u^\mathrm{o}t$ and $u't$), whereupon use of expressions such as eqns (2.43) and (2.44) would not be feasible for the evaluation of equilibrium constants from experimental patterns.

Although patterns of the type shown in Fig. 5.8 obviously bear much closer resemblance to experimental profiles than do their asymptotic counterparts, the absence of flow interactions (Baldwin, Dunlop, and Gosting 1955, Dunlop and Gosting 1955) has, of course, been assumed. To this end, Cox (1969) has outlined a procedure for the inclusion of any cross-term diffusion coefficients, but, unfortunately, evaluation of their magnitudes poses a far greater problem. Thus it may well be that instead of endeavouring to obtain theoretical patterns which emulate the experimental profiles, we should, perhaps, be attempting to obtain experimental results which resemble the asymptotic patterns. In this connection, Gilbert and his collaborators (Chiancone *et al.* 1968, Gilbert and Gilbert 1968) have already given the lead by noting that such patterns may in principle be constructed from the results of a series of differential plateau experiments in which $\bar{c}$ is either increased or decreased incrementally: the experimental feasibility of so doing is discussed and illustrated in section 3.4.

(c) *Boundary spreading in gel chromatography*

In the discussion of the countercurrent distribution analogue of velocity sedimentation (Bethune and Kegeles 1961*c*) it was noted that a major objection to this model is its incorrect prediction of the relative magnitudes of *s* and

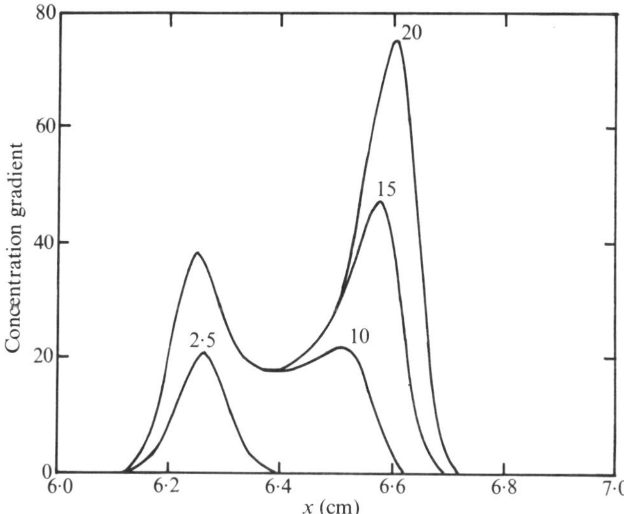

FIG. 5.8. Numerical examples illustrating the incorporation of diffusional spreading into theoretical sedimentation velocity patterns for a monomer–hexamer system. Magnitudes of parameters were as follows: rotor speed, 59 780 rev/min; meniscus position, 6·085 cm; $s_A = 6{\cdot}0\,S$, $s_C = s_A n^{2/3}$; $D_A = 8{\cdot}0 \times 10^{-7}\,\text{cm}^2\,\text{s}^{-1}$, $D_C = D_A n^{-1/3}$; $K = 1{\cdot}6 \times 10^{-5}\,\text{litre}^5\,\text{g}^{-5}$; $t = 1200\,\text{s}$; initial solute concentration ($\bar{c}_0$), 2·5–20 mg/ml as indicated. (Adapted from Cox (1969).)

D, because both are related to a single partition coefficient. However, it may well prove to be a better analogue of gel chromatography inasmuch as both processes involve partitioning of solute between two phases. Nevertheless, the discontinuous nature of the countercurrent (theoretical plate) model differs from an experimental system with continuous migration of mobile phase, and it is thus necessary to demonstrate that the analogue predicts correct boundary spreading for non-interacting solutes.

Implicit in the use of the countercurrent analogue is the assumption that diffusional and non-equilibrium effects can be ignored. In this connection the frequently observed insensitivity of elution profiles to column flow rate (for example, Flodin 1962, Ackers 1964, Siegel and Monty 1966, Hibberd *et al.* 1970) is evidence that a simple model based on partition equilibrium suffices as a satisfactory *operational* definition of gel chromatography, even though this insensitivity may well reflect the mutual cancellation of non-equilibrium and diffusion effects (Giddings and Mallik 1966). In the random-walk concept of chromatography, an equilibrium model, dispersion arises simply through granularity of the column and the consequent tortuosity of the paths followed by solute molecules during migration. Two versions of this model are possible (Beran 1957, Hibberd *et al.* 1970), the former (Beran 1957) predicting Gaussian distributions within the column (at a given time) and, therefore, non-Gaussian elution profiles: the reverse conclusion results from the latter treatment (Hibberd *et al.* 1970). This discrepancy reflects the fact that Beran has assumed that the elementary axial distances travelled by individual solute molecules in a fixed time increment are dispersed about a mean, whereas the alternative treatment considers that the times (τ) taken to migrate a given axial distance (λ) are so distributed, with root-mean-square dispersion $\alpha\tau$: either assumption seems plausible theoretically. Hibberd *et al.* (1970) were unable to detect skewness of frontal elution profiles for single, non-interacting solutes, a result confirmed by Halvorson and Ackers (1971), who, however, demonstrated asymmetry of zonal profiles. Although tests of skewness are more sensitive for zonal patterns, the assumptions about the original boundary conditions are also more suspect. The standard deviation of the distribution of elution volume (σ_{V_e}) is given by eqn (5.29),

$$\sigma_{V_e} = \frac{\alpha \bar{V}_e}{\sqrt{n}}, \tag{5.29}$$

where $\bar{V}_e$ is the mean elution volume and n denotes the number of incremental axial distances λ comprising the total column length: Fig. 4 of Hibberd *et al.* (1970), which summarizes gel chromatographic results obtained with a series of purified proteins on Sephadex G-100, provides a satisfactory test of this predicted linearity between σ_{V_e} and $\bar{V}_e$.

Of the various theoretical plate treatments of chromatography, the countercurrent model with continuous flow of mobile phase (Glueckauf 1955) resembles most closely the experimental situation. For a column comprising a large number of theoretical plates (N_p) this treatment leads to a Guassian distribution of solute within the column at any given time, the standard deviation, σ, at time t corresponding to $\bar{V}_e$ being given by

$$\sigma^2 = N_p(\Delta l)^2 = \frac{l^2}{N_p}, \tag{5.30}$$

Δl being the plate length ($=l/N_p$). If, as an approximation, the distribution on the column and the elution profile are both considered to be Gaussian, the standard deviation of the profile, σ_{V_e}, corresponding to σ on the column is (Hibberd *et al.* 1970)

$$\sigma_{V_e} = \frac{\bar{V}_e}{\sqrt{N_p}}, \tag{5.31}$$

which is of exactly the same form as eqn (5.29) predicted by random-walk theory: thus the slope of the linear $\sigma_{V_e} - \bar{V}_e$ plot may be used to define N_p, the number of theoretical plates comprising the column.

The computer programme for the Bethune and Kegeles (1961*a–c*) countercurrent analogue of free migration is readily adapted so that it approximates to the Glueckauf (1955) theoretical plate model of chromatography. Modifications include (i) transfer of volume increments (δV) smaller than the mobile phase volume of a single plate (V_p) and (ii) recording of the composition of the mobile phase being removed from the final plate at each transfer. The former change, which clearly brings the model more into line with the Glueckauf continuous countercurrent theory, is essential, since the Bethune and Kegeles analogue predicts proportionality between σ_{V_e} and $\bar{V}_e^{1/2}$ (cf. eqns (5.29) and (5.31)): a value of 0·1 for the ratio $\delta V/V_p$ has yielded estimates of σ_{V_e} approaching very closely those predicted (eqn (5.31)) for the countercurrent model with continuous flow. The second modification is dictated by the nature of the experimental record that is usually obtained (an elution profile instead of a *c–x* distribution). Non-Gaussian profiles are, of course, predicted for non-interacting solutes by this analogue, the estimates of σ_{V_e} from the upper and lower portions of simulated boundaries differing by approximately 5 per cent. With these adaptations the countercurrent treatment provides a useful, though perhaps not perfect, analogue of the actual chromatographic process.

5.5. Frictional interactions

(a) *Effects in* systems *comprising one constituent*

Consider first the mass migration of a single solute which is not involved in any chemical reaction. The continuity equation (eqn (2.5)) must apply. If v is a constant, independent of total concentration, we have seen that this equation describes (in the absence of diffusional spreading) a sharp boundary between solvent and solution plateaux: the conjugate ascending and descending patterns are enantiographic. However, when the velocity of the solute exhibits a dependence on total concentration due to frictional or physical interaction (a common phenomenon in sedimentation velocity and gel chromatography of macromolecular solutes), this simple interpretation of eqn (2.5) is no longer possible even though eqn (2.5) written in terms of constituent quantities continues to express correctly the conservation of mass in the absence of diffusional flows. In the event that frictional interactions operate (i.e. $\mathrm{d}\bar{v}/\mathrm{d}\bar{c} \neq 0$) eqn (2.5) may only be expanded in the form

$$(u-\bar{v})\,\mathrm{d}\bar{c}-\bar{c}\,\mathrm{d}\bar{v}=0, \tag{5.32}$$

which may be differentiated with respect to $\bar{c}$ to obtain

$$\frac{\mathrm{d}u}{\mathrm{d}\bar{c}}=2\frac{\mathrm{d}\bar{v}}{\mathrm{d}\bar{c}}+\bar{c}\frac{\mathrm{d}^2\bar{v}}{\mathrm{d}\bar{c}^2}. \tag{5.33}$$

If the convention is adopted that distance, x, is positive in the direction solvent$\rightarrow$solution, $\mathrm{d}\bar{c}/\mathrm{d}x$ must be positive in both ascending and descending sides. In the descending side u is positive by definition of x, whereupon the boundary is hypersharp (Nichol and Ogston 1965*a*) if u decreases with increasing $\bar{c}(\mathrm{d}\bar{c}/\mathrm{d}u<0)$. For example, a linear negative dependence of sedimentation coefficient on total solute concentration is frequently observed in ultracentrifugation of globular proteins, such that

$$\bar{v}=v^{\circ}(1-b\bar{c}), \tag{5.34}$$

where v° and b are positive constants for the descending side. It follows from eqn (5.34) that $\mathrm{d}\bar{v}/\mathrm{d}\bar{c}=-bv^{\circ}<0$ and that $\mathrm{d}^2\bar{v}/\mathrm{d}\bar{c}^2=0$: consequently, from eqn (5.33), $\mathrm{d}u/\mathrm{d}\bar{c}<0$. The hypersharp boundary moves with a unique velocity at a specified concentration described by eqn (3.4). Thus, provided we start with an initially sharp boundary at $t=0$, the hypersharp boundary is formed immediately and travels with constant velocity (Nichol *et al.* 1970). In practice diffusional spreading will oppose this tendency to form a hypersharp boundary, and conditions may be selected which lead to a steady state (Creeth 1964). For the example specified in eqn (5.34) all velocities are negative on the conjugate ascending side (because of the convention adopted to define x), but

b is still positive. By considering the first and second derivatives of $\bar{v}$ with respect to $\bar{c}$ we find that $\mathrm{d}u/\mathrm{d}\bar{c}$ is positive, and thus a spread ascending boundary is predicted. By combining eqns (5.34) and (5.32) we obtain

$$(u - v^{\mathrm{o}} + 2bv^{\mathrm{o}}\bar{c})\ \mathrm{d}\bar{c} = 0. \tag{5.35}$$

Thus either $\mathrm{d}\bar{c} = 0$, which describes the solvent and solution plateaux, or the term within parentheses is zero, which describes the spread boundary, viz.

$$u = v^{\mathrm{o}} - 2bv^{\mathrm{o}}\bar{c}. \tag{5.36}$$

Obviously, in the boundary region the plot of $\bar{c}$ vs. u is linear, while in the derivative plot $\mathrm{d}\bar{c}/\mathrm{d}u$ maintains a constant value of $-1/(2bv^{\mathrm{o}})$ for all values of u in the boundary region, it being zero elsewhere. Linear concentration-dependence of velocities does not, therefore, introduce asymmetry of boundaries in either ascending or descending sides (cf. Fujita 1956).

A linear concentration dependence of elution volume on total concentration has been observed in frontal gel chromatography of ovalbumin (Winzor and Nichol 1965), the elution volume increasing with total concentration, behaviour which contrasts with that of the associating α-chymotrypsin system (Winzor and Scheraga 1964). The dependence observed with ovalbumin was accordingly attributed to a physical interaction inasmuch as there was a dependence of the partition coefficient, describing the distribution of the protein between mobile and gel phases, on the composition of mobile solution in which no chemical equilibria were operating (Nichol *et al.* 1967*b*). Elution profiles obtained with ovalbumin exhibited symmetrical ascending and descending boundaries, separated by a plateau of original composition: in accord with the theoretical prediction the ascending boundary spread considerably more than its descending counterpart. The overall asymmetry of the complete elution profile (Fig. 5.9) is thus a consequence of the recording of both ascending and descending profiles: similarly, concentration-dependent migration leads to asymmetry of zones in which no plateau region exists (DeVault 1943, Houghton 1963). It is clear, however, that the method of frontal analysis provides a distinction between the effects of heterogeneity and linear concentration dependence in that neither ascending nor descending sides are symmetrical for a paucidisperse system (for example, Fig. 2.1). In addition, it permits the evaluation of linear $\bar{v}$–$\bar{c}$ dependence from a single plateau experiment (Ronalds and Winzor 1969).

It is important to recall that eqn (5.34) expressed only one possible relationship between the constituent velocity of a single solute and its total concentration, and that the conclusions drawn concerning a hypersharp boundary on the descending side and a spread ascending boundary are, therefore, not general. It is possible, for example, that on the descending side (where $d\bar{c}/dx$ and u are both positive according to the convention adopted about the sign of x) that u could increase with $\bar{c}$ ($d\bar{c}/du > 0$), leading to a spread boundary on the descending side. Moreover, the boundary would be symmetrical only in the event that $\bar{v}$ varied linearly with $\bar{c}$. It is also clear from eqn (5.33) that an increase of $\bar{v}$ with $\bar{c}$ ($d\bar{v}/d\bar{c} > 0$) is not a sufficient condition to ensure that a boundary will spread ($d\bar{c}/du > 0$): in eqn (5.33) $\bar{c}$ is necessarily positive but the second derivative may have either positive or negative sign. Indeed, with certain functional relationships between $\bar{v}$ and $\bar{c}$ it is possible for the right

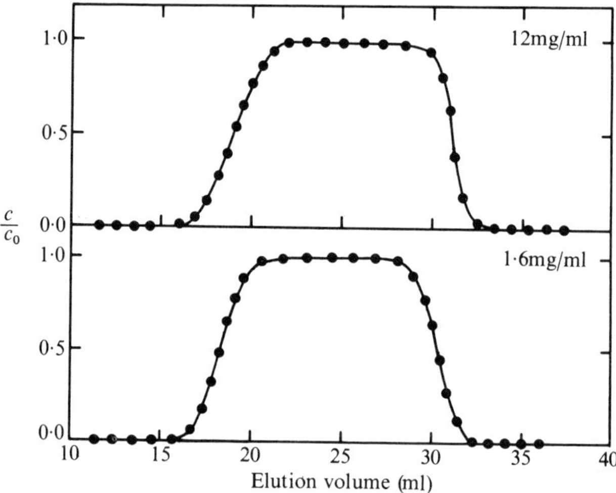

FIG. 5.9. Effects of linear concentration dependence of elution volume upon the protein distribution in gel chromatographic experiments on 12 ml zones of ovalbumin on a 1·1 × 32 cm column of Sephadex G-100 equilibrated with 0·1 I phosphate, pH 6·8. (Adapted from Winzor and Nichol (1965).)

hand side of eqn (5.33) to be positive within a range of $\bar{c}$ and negative outside of it. This means that the boundary will consist in part of a spread region ($d\bar{c}/du > 0$) and in part of a hypersharp region ($d\bar{c}/du < 0$). Nichol and Ogston (1965*a*) illustrated this type of behaviour by selecting arbitrarily the following relations describing the dependence of $\bar{v}$ on $\bar{c}$:

$$\bar{v} = v_A + v(1 - e^{-b\bar{c}}), \tag{5.37a}$$

$$\bar{v} = v_A + \frac{vb\bar{c}^n}{(1 + b\bar{c}^n)}, \tag{5.37b}$$

where v_A, v, b, and n are constants. They also drew the general conclusion that if one boundary or part of a boundary spreads, the conjugate corresponding boundary or part thereof must be hypersharp, unless conditions differ between the two sides other than with respect to the change of signs required by the definition of x. This type of finding may, at first sight, appear to be of academic interest only. However, it assumes practical importance when it is recalled from Chapter 3 that the sedimentation of a polymerizing system ($n\mathrm{A} \rightleftharpoons \mathrm{C}$, $n > 1$, $v_\mathrm{C} > v_\mathrm{A}$) is expected to give rise to a spread reaction boundary, whereas linear negative concentration-dependence of both species velocities would tend to introduce hypersharpening of the same boundary.

Gilbert (1963) treated this problem of a frictional variation in migration rate superimposed upon a polymerization equilibrium. For each ith polymer species he formulated an expression entirely analogous to eqn (5.34) to account for frictional effects, viz.

$$v_i = v_i^\mathrm{o}(1 - b\bar{c}), \tag{5.38}$$

where the coefficient b is considered to be identical for all species, and $\bar{c}$ is the constituent concentration. Whereas in the absence of frictional interaction ($b=0$) the dependence of the constituent velocity on constituent concentration was described by the law of mass action according to eqn (3.14b), the operation of frictional effects requires reformulation of the dependence as

$$\bar{v}\bar{c} = M_{\mathrm{A}_1} \sum_{i=1}^{i=n} i v_i^\mathrm{o}(1 - b\bar{c}) m_{\mathrm{A}_i}. \tag{5.39}$$

Suppose now that we wish to construct a numerical plot of $\bar{c}$ vs. u for a chosen set of the parameters M_{A_1}, n, K_i (equilibrium constants defined in eqn (3.13)), v_i^o, and b. For a range of values of m_{A_1}, the molar concentration of monomer, corresponding values of $\bar{c}$ may readily be calculated from eqn (3.14a), and it therefore remains to evaluate the corresponding values of u. For an initially sharp boundary eqn (3.15a) applies, as does eqn (3.15c); but the expression for Q must be reformulated since $M_{\mathrm{A}_1} Q = \mathrm{d}\bar{v}\bar{c}/\mathrm{d}m_{\mathrm{A}_1}$. Differentiation of eqn (5.39) with respect to m_{A_1} and collection of terms gives

$$Q = (1 - b\bar{c}) \sum_{i=1}^{i=n} i v_i^\mathrm{o} \frac{\mathrm{d}m_{\mathrm{A}_i}}{\mathrm{d}m_{\mathrm{A}_1}} - \sum_{i=1}^{i=n} i v_i^\mathrm{o} b m_{\mathrm{A}_i} \frac{\mathrm{d}\bar{c}}{\mathrm{d}m_{\mathrm{A}_1}}. \tag{5.40}$$

From eqn (5.40) we may define $Q = Q_\mathrm{o}$ when $b=0$, whereupon Q_o may be evaluated for the range of values of m_{A_1} according to the numerator of eqn (3.15b), noting that when $b=0$, v_i and v_i^o are identical (eqn (5.38)). If it is also recalled that $M_{\mathrm{A}_1} P = \mathrm{d}\bar{c}/\mathrm{d}m_{\mathrm{A}_1}$ and that P may be evaluated from the denominator of eqn (3.15b), we find on combination of eqns (3.15c) and (5.40) that

$$u = \frac{Q}{P} = \frac{(1 - b\bar{c})Q_\mathrm{o}}{P} - M_{\mathrm{A}_1} b \sum_{i=1}^{i=n} i v_i^\mathrm{o} K_i (m_{\mathrm{A}_1})^i. \tag{5.41}$$

Thus corresponding values of u and $\bar{c}$ may be computed for a selected range of values of m_{A_1}. Gilbert (1963) also formulated the differential expression (eqn (3.17b)) in terms of the present model. His approach elegantly illustrates the feasibility of introducing frictional relations, such as eqn (5.38) which is, of course, only one possible relation, into his initial theory designed to account for interactions governed by the law of mass action. Such allowance for effects of physical interactions on migration patterns for reversibly polymerizing systems is to be found, for example, in the interpretation of velocity sedimentation results on β-lactoglobulin (Gilbert 1963) and of gel chromatographic data on haemoglobin (Chiancone *et al.* 1968) and α-chymotrypsin (Ackers 1967).

(b) *Effects in systems comprising two or more constituents*

Davies, Nichol, and Ogston (1963) pointed out that the conceptual elegance of the constituent approach is that it predicts the formation of reaction boundaries (spread and hypersharp) when *any* type of interaction modifies the velocities of the migrating species and thereby renders the constituent velocities dependent on solution composition. We have seen in previous chapters illustrations of this prediction for systems involving chemical interactions, and in the above section on single solutes, examples wherein frictional interactions, either alone or in combination with chemical equilibria, operate. We now consider the migration of two constituents A and B, the relevant constituent velocities being written in general form as

$$\bar{v}_A = v_A + f(\bar{m}_A, \bar{m}_B), \qquad f(0, 0) = 0, \tag{5.42a}$$

$$\bar{v}_B = v_B + \phi(\bar{m}_A, \bar{m}_B), \qquad \phi(0, 0) = 0. \tag{5.42b}$$

The causes of the composition-dependent changes in constituent velocities have not been specified, but it has been assumed that at zero concentration of both species the limiting values of the constituent velocities are v_A and v_B. Combination of eqns (5.42a, b) with the continuity equations (eqns (2.11a, b)) gives

$$(u_A - v_A - f)\, d\bar{m}_A - \bar{m}_A\, df = 0, \tag{5.43a}$$

$$(u_B - v_B - \phi)\, d\bar{m}_B - \bar{m}_B\, d\phi = 0. \tag{5.43b}$$

Partial differentiation of eqns (5.42a, b) permits eqns (5.43a, b) to be rewritten as

$$\left(u_A - v_A - f - \bar{m}_A \frac{\partial f}{\partial \bar{m}_A}\right) d\bar{m}_A - \bar{m}_A \frac{\partial f}{\partial \bar{m}_B}\, d\bar{m}_B = 0, \tag{5.44a}$$

$$\left(u_B - v_B - \phi - \bar{m}_B \frac{\partial \phi}{\partial \bar{m}_B}\right) d\bar{m}_B - \bar{m}_B \frac{\partial \phi}{\partial \bar{m}_A}\, d\bar{m}_A = 0. \tag{5.44b}$$

When $d\bar{m}_A=0$, there is a plateau region of A wherein *either* $\bar{m}_A \neq 0$ *or* $\bar{m}_A=0$. If $\bar{m}_A \neq 0$, eqn (5.44a) shows that $d\bar{m}_B=0$, which means that any plateau of A is also a plateau of B. If $\bar{m}_A=0$, eqn (5.44b) becomes

$$u_B - v_B - \phi(0, \bar{m}_B) - \bar{m}_B \frac{\partial \phi(0, \bar{m}_B)}{\partial \bar{m}_B} d\bar{m}_B = 0, \tag{5.45}$$

which has two solutions. One is that $d\bar{m}_B=0$, which describes a plateau of B wherein either $\bar{m}_B=0$ or $\bar{m}_B \neq 0$. To prevent repetition of argument let us assume that B is the more slowly migrating material, in which case the solution $d\bar{m}_B=0$, $\bar{m}_B=0$ describes the solvent plateau. The solution $d\bar{m}_B=0$, $\bar{m}_B \neq 0$ then describes the finite plateau of B between a boundary of pure B and the position where $\bar{m}_A$ begins to increase from zero. The other solution of eqn (5.45), applicable when $\bar{m}_A=0$ and $d\bar{m}_B \neq 0$ (noting that $\bar{m}_B=m_B$ under these circumstances) is

$$u_B - v_B - \phi(0, m_B) - m_B \frac{\partial \phi(0, m_B)}{\partial m_B} = 0, \tag{5.46}$$

which describes the boundary of pure B. Obviously, if $\phi(0, m_B)=0$ it would be a sharp boundary at the position $v_B t$. However, depending on the form of $\phi(0, m_B)$ it may be spread or hypersharp or partially both. Indeed, in special cases (for example, Fig. 4.2) there may be no boundary due to pure solute (A or B), a situation which is also consistent with the present general formulation. Clearly, this formulation can be extended only in the event that the functions $\partial f/\partial \bar{m}_A$, $\partial f/\partial \bar{m}_B$, $\partial \phi/\partial \bar{m}_A$, and $\partial \phi/\partial \bar{m}_B$ are given explicit form. However, the approach suffices to show that the *general character* of composition distribution in a migration system is independent of the exact forms of the functions of composition.

Accordingly, let us consider a boundary distribution such as that described above, in which a boundary of pure B arises from a solvent plateau (γ phase) to a plateau containing B alone (β phase), which is terminated at a position where $\bar{m}_A$ begins to increase. This distribution may be summarized, using the nomenclature of Longsworth (1959*a*), as

$$\text{S, R, A, B}(\alpha) \leftarrow \text{S, R, B}(\beta) \leftarrow \text{S, R}(\gamma), \tag{5.47}$$

where arrows represent moving boundaries, S and R anions and cations of the supporting electrolyte, and the α-phase is the plateau region of original solution composition. Since A and B are being considered as constituents here, it is understood that the α phase contains (in addition to supporting electrolyte) the *solute* species A and B (if no chemical reaction occurs) or, for example A, B, and C (if a reaction $A+B \rightleftharpoons C$ occurs). In connection with this distribution, one question of considerable interest pertains to the relation between m_B^β and $\bar{m}_B^\alpha$. Nichol and Ogston (1965*b*) examined this question by applying simple geometry to a model (Fig. 5.10), which illustrates possible distributions of the *constituent* concentration of B at two times, $t=0$ and $t=t$: it was shown that no generality was lost by considering this particular distribution. At $t=0$ a sharp boundary is present at the position AB, whereupon the initial, total amount of B in any form between this plane and plane JM is given by the area ABJM (the 'rectangular approximation' being made). After applying a homogeneous field for time t, the distribution of the B constituent is

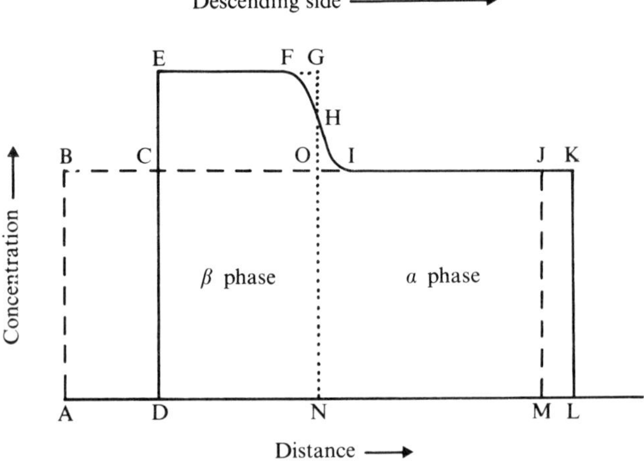

FIG. 5.10. A representative distribution of the *constituent* concentration of slower constituent (B) in a descending limb at two different times, viz. $t=0$ (– – –) and $t=t$ (———). The gradient in constituent concentration of A, which also separates the α and β phases, has been omitted for simplicity. (Adapted from Nichol and Ogston (1965*b*).)

shown by the solid line DEFIKL. If the plane JM represented the base of the cell in sedimentation velocity, the quantity of B corresponding to the area MJKL (extending hypothetically beyond the cell bottom) would have piled up at JM. Conservation of matter requires that

$$\text{area ABJM} = \text{area DEFIKL}. \tag{5.48}$$

The gradient of concentration of the B constituent in the $\alpha\beta$ region has been drawn arbitrarily as a spread negative gradient at time t. A vertical line GHON is constructed perpendicular to the distance ordinate and bisecting the curve FI at the median position H, such that the areas FGH and OHI are equal. It follows that

$$\text{area DEFIKL} = \text{area CEGO} + \text{area DCJM} + \text{area MJKL}. \tag{5.49}$$

It is important to note that eqn (5.48) is valid whether the gradient FI is spread and negative (as shown in Fig. 5.10), is sharp or hypersharp, or is spread and positive. It is also noted that the line ECD may be regarded as the median of a curve describing a diffuse boundary of B. Clearly,

$$\text{area ABJM} = \text{area ABCD} + \text{area DCJM}. \tag{5.50}$$

Combining eqns (5.48)–(5.50) we obtain

$$\text{area ABCD} = \text{area CEGO} + \text{area MJKL}. \tag{5.51}$$

Equation (5.51) may be rewritten in terms of concentrations and velocities on noting the following: $AB = \bar{m}_B^\alpha = JM$; $CE = (m_B^\beta - \bar{m}_B^\alpha)$; $AD = v_B^\beta t$; $ML = \bar{v}_B^\alpha t$; $CO = (v' - v_B^\beta)t$, where v' is the velocity of the median GHON. With these substitutions, eqn (5.51) becomes

$$\bar{m}_B^\alpha v_B^\beta t = (m_B^\beta - \bar{m}_B^\alpha)(v' - v_B^\beta)t + \bar{m}_B^\alpha \bar{v}_B^\alpha t \tag{5.52a}$$

or

$$m_B^\beta = \bar{m}_B^\alpha \left(\frac{v' - \bar{v}_B^\alpha}{v' - v_B^\beta} \right). \tag{5.52b}$$

Equation (5.52b) is the integrated form of eqn (2.11b); but our patience with the geometrical argument has been rewarded by recognizing that v' is the velocity of the median bisector of the concentration gradient of *constituent* B in the region of the $\alpha\beta$ boundary, across which constituent A disappears.

Equation (5.52b) may be viewed in several different ways. First, if the states of B in the β and α regions are equated with hydrodynamically unperturbed B and B frictionally perturbed by A, respectively (i.e. no chemical reaction occurs), eqn (5.52b) is the 'Johnston–Ogston equation' (Johnston and Ogston 1946): these workers, however, failed to identify v' as defined above. In the case of frictional interaction between two solute species it is generally found (at least in sedimentation velocity) that $v_B^\beta > \bar{v}_B^\alpha$ and hence (from eqn (5.52b)) that $m_B^\beta > \bar{m}_B^\alpha$, the situation illustrated in Fig. 5.10. That neglect of this effect may lead to incorrect quantitative analyses of mixtures by mass migration methods has been amply stressed in relation to sedimentation velocity studies (for example, Schachman 1959); but the effect has also been shown to operate in frontal gel chromatographic experiments performed with mixtures of macromolecular solutes which do not interact *chemically* (Winzor and Nichol 1965). Secondly, if the equilibrium reaction $A + B \rightleftharpoons C$ occurs but no frictional effects operate (i.e. v_A, v_B, and v_C are constants, whereupon $v_B^\beta = v_B^\alpha$), eqn (5.52b) becomes the 'moving boundary equation' used by Longsworth (1959*a*), Alberty and Marvin (1950) and Nichol and Winzor (1964). In the event that $v_A > v_B$ on the descending side complex formation ensures that $\bar{v}_B^\alpha > v_B^\beta$, and, from eqn (5.52b), that $\bar{m}_B^\alpha > m_B^\beta$. However, as noted in Chapter 4 in relation to

Fig. 2.4, a descending system for the case $v_A > v_B = v_C$ with the slower reactant labelled B is characterized by the unique relation $v_B^\beta = \bar{v}_B^\alpha$, whence, from eqn (5.52b), $m_B^\beta = \bar{m}_B^\alpha$: this conclusion holds only in the absence of frictional interactions. Finally, eqn (5.52b) may be employed as a guide to the possible relative magnitudes of m_B^β and $\bar{m}_B^\alpha$ when both chemical ($A + B \rightleftharpoons C$) and frictional effects are reflected in the boundary distribution under discussion (condition (5.47)). The possible diverse behaviour is summarized in Table 5.1, which has been adapted from Nichol and Ogston (1965*a*).

TABLE 5.1

A comparison of the magnitudes of m_B^β and $\bar{m}_B^\alpha$ for the distribution S, R, A, B, C(α)←S, R, B(β)←S, R(γ)

Case	Relative magnitudes
(1) $v_C^\alpha < v_B^\alpha$, i.e. $v_B^\alpha > \bar{v}_B^\alpha$	
(a) $v_B^\beta > v_B^\alpha$	$m_B^\beta > \bar{m}_B^\alpha$
(b) $v_B^\beta < v_B^\alpha$	$m_B^\beta \geq \bar{m}_B^\alpha$ or $m_B^\beta \leq \bar{m}_B^\alpha$
(c) $v_B^\beta = v_B^\alpha$	$m_B^\beta > \bar{m}_B^\alpha$
(2) $v_C^\alpha > v_B^\alpha$, i.e. $v_B^\alpha < \bar{v}_B^\alpha$	
(a) $v_B^\beta > v_B^\alpha$	$m_B^\beta \geq \bar{m}_B^\alpha$ or $m_B^\beta \leq \bar{m}_B^\alpha$
(b) $v_B^\beta < v_B^\alpha$	$m_B^\beta < \bar{m}_B^\alpha$
(c) $v_B^\beta = v_B^\alpha$	$m_B^\beta < \bar{m}_B^\alpha$
(3) $v_C^\alpha = v_B^\alpha$, i.e. $v_B^\alpha = \bar{v}_B^\alpha$	
(a) $v_B^\beta > v_B^\alpha$	$m_B^\beta > \bar{m}_B^\alpha$
(b) $v_B^\beta < v_B^\alpha$	$m_B^\beta < \bar{m}_B^\alpha$
(c) $v_B^\beta = v_B^\alpha$	$m_B^\beta = \bar{m}_B^\alpha$

While it is apparent that the complete elucidation of the distribution specified in condition (5.47) must await specification of the functions f and ϕ (eqns (5.42a, b) in the event that both chemical and physical interactions operate, this realization is of little help to the experimenter confronted with the problem of interpreting such a distribution. The detailed problems may be enunciated more clearly with a particular example in which migration of a two-constituent system undergoing the chemical reaction $A + B \rightleftharpoons C$ results in a distribution given by condition (5.47), with frictional perturbations of velocities also occurring in the α phase. We do not suggest that the specified distributions must necessarily arise in the study of such a system (exceptions have been noted in Chapter 4); but such a distribution is possible, and may be confirmed experimentally by sampling procedures. In such a situation expressions for m_B^α and $\bar{m}_B^\alpha \bar{v}_B^\alpha$ obtained by rearrangement of eqns (2.9a, b, c, d), viz.

$$m_B^\alpha = \bar{m}_B^\alpha - \bar{m}_A^\alpha + m_A^\alpha \qquad (5.53a)$$

$$\bar{m}_B^\alpha \bar{v}_B^\alpha = m_B^\alpha v_B^\alpha + \bar{m}_A^\alpha \bar{v}_A^\alpha - m_A^\alpha v_A^\alpha \qquad (5.53b)$$

may be substituted into eqn (5.52b) to give

$$m_A^\alpha(v_A^\alpha - v_B^\alpha) = \bar{m}_A^\alpha(\bar{v}_A^\alpha - v_B^\alpha) - \bar{m}_B^\alpha(v' - v_B^\alpha) + m_B^\beta(v' - v_B^\beta). \qquad (5.54)$$

Application of eqn (5.54), which holds only for a cell of uniform cross-section, to experimental profiles requires knowledge of v_A^α and v_B^α, velocities which reflect the effects of mutual physical interactions between reactant and product species, and which may not, therefore, be obtainable from separate studies of the individual reactants. The quantities $\bar{m}_A^\alpha$ and $\bar{m}_B^\alpha$ may be obtained from the concentrations of A and B mixed initially, while m_B^β and v_B^β are found directly from the slow moving boundary on the descending side. Provided the parameters $\bar{v}_A^\alpha$, v', v_A^α, and v_B^α may be *estimated*, eqn (5.54) yields m_A^α, whereupon K may be obtained from eqn (5.55).

$$K = \frac{m_C^\alpha}{m_A^\alpha m_B^\alpha} = \frac{(\bar{m}_A^\alpha - m_A^\alpha)}{m_A^\alpha(\bar{m}_B^\alpha - \bar{m}_A^\alpha + m_A^\alpha)}. \qquad (5.55)$$

Consider first the estimation of v', the velocity of the median bisector of the B constituent gradient in the $\alpha\beta$ region. Clearly, exact definition of v' requires a record of the sum of the concentrations of complexed and free B, which may be difficult even with spectral measurements because of the joint presence of A and also of a possible spectral shift accompanying the formation of C. Indeed, in many practical situations the experimental record does not distinguish between the constituents present, it frequently being a plot of the total concentration (absorbance) of all constituents vs. distance (volume) or a curve of the total refractive index gradient vs. distance (as in a schlieren pattern). One approach, and it must be regarded as approximate (Gilbert and Kellett 1971), is to assume that v' and $\bar{v}_A^\alpha$ are identical, both being given by the median bisector of the total concentration gradient across the $\alpha\beta$ boundary. There still remains the problem of estimating v_A^α and v_B^α. Basically two approaches, both approximate, seem possible. The first involves a study of a mixture of A and B in an environment where chemical reaction is prevented (for example, at a different pH): provided $\bar{m}_A^\alpha$ and $\bar{m}_B^\alpha$ are unchanged from the values used under conditions of complex formation, estimates of v_A^α and v_B^α may bc obtained from the appropriate descending and ascending boundaries, respectively, using again the median bisector of the total concentration gradient. The second approach involves the determination of the individual concentration-dependences of v_A and v_B in the environment used to study the chemically reacting mixture, and using as estimates of v_A^α and v_B^α the values of v_A and v_B corresponding to (say) a weight-concentration equivalent to $(\bar{m}_A + \bar{m}_B)$ in the mixture. Nichol, Ogston, and Winzor (1967*c*), who formulated eqn (5.54) in terms of elution volumes, discuss the joint application of both approaches in a reiterative procedure designed to evaluate v_A^α and v_B^α, thence K, from frontal gel chromatographic results obtained with mixtures of ovalbumin and lysozyme under conditions favouring complex formation by electrostatic interaction.

6

CONCLUDING REMARKS

IT has become clear that any type of interaction which conveys a dependence of constituent velocities on composition or time will result in complicated mass migration patterns. The ability to interpret these patterns involving spread and/or hypersharp reaction boundaries relies on the availability of explicit expressions describing constituent velocities as a function of the relevant variables. In some cases, notably those involving frictional interactions, these expressions are lacking at present, but the potential for studying such effects by mass migration methods remains and should be stressed, especially since equilibrium methods cannot contribute directly in this field. In this review, emphasis has been given to rapidly-established chemical equilibria for two main reasons. First, the required dependence of constituent velocities on composition is provided by the law of mass action, and, secondly, acceptor-ligand interactions of this type are of wide occurrence. The detection of these interactions is a relatively simple matter (except, ironically, for the simplest isomerization reaction). In general, mass migration patterns for these systems will exhibit non-enantiography of conjugate sides separated by a plateau region of original composition, variation of the total concentration (of a single associating solute) or composition (of a mixture of interacting solutes) in the original plateau leading to changes in the experimental pattern that are not predictable by classical theories nor necessarily by physical reasoning. An illustration is provided by the unique concentration-dependence of the area distribution of a bimodal reaction boundary observed in the migration of a system comprising monomer in rapidly established equilibrium with a single higher polymer. However, even in this case, we recommend that quantitative information on the stoichiometry and equilibrium constant be obtained from the concentration-dependence of weight- and z-average velocities (or elution volumes) derived by integrating results across the *entire* boundary region, at all points of which monomer and polymer coexist in equilibrium. In this connection, we would agree basically with those who claim that the corresponding treatment of average molecular weights, derived from light scattering, osmotic pressure, and sedimentation equilibrium, offer a superior way of characterizing systems of the type $n\mathrm{A} \rightleftharpoons \mathrm{C}$ and $m\mathrm{A} + n\mathrm{B} \rightleftharpoons \mathrm{C}$ (rapidly established or kinetically controlled), especially since molecular weights of complex species bear integral relation to those of reactants, unlike the corresponding velocity terms. At the same time, we note the unique advantage of frontal gel chromatography for following (by enzymic assay) the behaviour of an interacting enzyme at high dilution, and also its use as a differential method for detecting small differences in (say) the extent of polymerization of variants of the same protein.

Frontal analysis in gel chromatography also offers a means of obtaining binding results with acceptor–ligand systems too labile to be subjected to prolonged equilibrium dialysis. We mention gel chromatography deliberately in this connection, since the relative order of magnitudes of the velocities of the individual species may be decided by the experimenter to ensure that the equilibrium concentration of unbound ligand is determined (Chapter 2). A major conclusion of the theoretical findings of Chapter 4 is that the concentration of free ligand is not obtained with systems wherein the velocities of complexes do not equal those of unbound forms of acceptor. It may be possible to employ the integrated moving boundary (flux, Johnston–Ogston) equation to correct ligand concentrations observed with certain of the other systems, but only if the experimenter appreciates the requirement of determining median bisectors of constituent gradients.

The various ideas and suggestions that we have attempted to develop coalesce when the interpretation of binding results is considered. Of the binding curves discussed, those sigmoidal in form are probably of the greatest biological interest. While it is legitimate and useful to replot sigmoidal binding results to produce non-linear double-reciprocal, Scatchard, or Hill plots, it is illogical to ascribe the non-linearity to mutually interacting sites (i.e. to a cooperative effect) without supporting evidence. Certainly such effects may operate in some systems, but it is also true that preferential binding of ligand to one form of an acceptor coexisting in equilibrium with a less active form gives rise to sigmoidality, even though binding sites within an acceptor entity are *equivalent and independent*. With these alternatives, the only recourse is to study experimentally the acceptor and ligand, first separately and then together, over a wide range of environmental conditions. Many techniques will be required, and we hope that this treatment of mass migration procedures may contribute to an overall understanding of these fascinating systems.

GLOSSARY OF SYMBOLS

The symbols A, B, C, D, E, S, and T are used to denote individual molecular species, and *i*, *j*, *k*, *l*, *m*, and *n* stoichiometry factors which describe the composition of any chemical complex of the individual species. In what follows, the symbol X is used as a subscript in reference to any of these species or complexes of them.

b	coefficient expressing concentration dependence of a velocity due to frictional interaction
c	concentration of a single, non-interacting solute (g/unit volume)
$c_{o,X}$	concentration of X in original solution plateau (g/unit volume)
c_X	concentration of species X (g/unit volume)
$\bar{c}$	constituent concentration of a single, self-interacting solute (g/unit volume)
$\bar{c}_o$	constituent concentration of single solute in original solution plateau
$\bar{c}_{min}$	constituent concentration at the position of the minimum in a bimodal reaction boundary
$\bar{c}_E$, $\bar{c}_L$, $\bar{c}_P$	total concentrations of effector, ligand, and acceptor, respectively (g/unit volume)
$\bar{c}_X$	constituent concentration of solute X within the stationary phase of a chromatography column
$\bar{\bar{c}}$	inclusive constituent concentration pertaining to chromatography
D_X	diffusion coefficient of species X
$\bar{D}$	average diffusion coefficient
F	number of classes of non-equivalent binding sites
G	stationary phase in chromatography
H	enthalpy
I	ionic strength
k	dissociation constant
k_{+1}, k_{-1}	rate constants
K	association equilibrium constant (g^{1-n} litre^{n-1} for $n\mathrm{A} \rightleftharpoons \mathrm{C}$; litre/mole for $\mathrm{A}+\mathrm{B} \rightleftharpoons \mathrm{C}$; dimensionless for $\mathrm{A}+\mathrm{B} \rightleftharpoons \mathrm{C}+\mathrm{D}$)
K_i	overall association equilibrium constant (M^{1-n}) for a polymerizing system ($i=1, 2, \ldots n$)
K'_i	step-wise association equilibrium constant (M^{-1}) for successive polymerization steps ($i=1, 2, \ldots n$)
K_X	intrinsic association binding constant of ligand to acceptor (X = A or C) (litre/mole)
l	length of column
L_X	intrinsic association binding constant of ligand to acceptor (X = A or C) (litre/mole)
m_X	concentration of species X (moles/litre)
$\bar{m}_X$	constituent concentration of X (moles/litre)
$\bar{\bar{m}}_X$	inclusive constituent concentration of X in chromatography
M	molarity
M_X	molecular weight of X
M_w	weight-average molecular weight
N_p	number of theoretical plates in a chromatography column

N_X	intrinsic association binding constant of effector to acceptor (X=A or C) (litre/mole)
p	number of binding sites per molecule of acceptor A
p_1, p_2	constituents of a hybrid species X
P	pressure
q	number of binding sites per molecule of acceptor A
q_1, q_2	constituents of a hybrid species X
r	binding function on a weight-scale
r_M	binding function on a molar scale
R	gas constant
s	sedimentation coefficient (s)
$\bar{s}$	weight-average sedimentation coefficient (s)
t	time
T	temperature
u	velocity of a lamina of constant composition ($=x/t$)
$u't$, $u^{o}t$	positions of termination of spread reaction boundaries
$\bar{u}t$	position of median bisector of a constituent gradient
$u_{min}t$	position of the minimum of a bimodal reaction boundary ($=x_{min}$)
U	electrophoretic mobility (cm^2 s^{-1} V^{-1})
v	velocity of migration of a single, non-interacting solute
v_X	velocity of migration of species X
$\bar{v}$, $\bar{v}_X$	constituent velocities of single species and X, respectively
$\bar{\bar{v}}$, $\bar{\bar{v}}_X$	inclusive constituent velocities pertaining to chromatography
v_{init}	initial velocity of a reaction
v_Δ	velocity of a differential boundary
V_X	elution volume of species X
$\bar{V}_X$	constituent elution volume of X
$\dot{V}$	flow rate of a column
V_l	loading volume on a column
V_o	void volume of a column
V', V^o	elution volumes corresponding to terminal points of a spread reaction boundary
V_{max}	maximal initial velocity of a reaction
V_w	weight-average elution volume
V_z	z-average elution volume
w	number of binding sites available to an effector
$w_{i,j}$	statistical weights of complexes
$w_{k,l}$	statistical weights of complexes
x	distance
x_m	distance of meniscus from centre of rotation in sedimentation
X	association equilibrium constant (M^{1-n}) for acceptor self-interaction (Chapter 1)
X'	apparent equilibrium constant in the presence of an effector
y	number of binding sites available to an effector
Y	association equilibrium constant (M^{1-n}) for ligand self-interaction (Chapter 1)
Y	as subscript denotes the general constituent p_1, p_2, q_1, or q_2
z	net charge (or valence)
Z	concentration gradient parameter
α	transform ($=K_A m_S$)

α_{A}	weight-fraction of monomer in a polymerizing system
α, β, γ	as superscripts denote phases or zones in cell (descending side)
β_j	number of binding sites on the jth acceptor
$\gamma_{\pm}$	mean ionic activity coefficient
η	viscosity of medium
Λ	transform $\Lambda = \{2v_{\mathrm{C}} - (v_{\mathrm{A}} + v_{\mathrm{B}})\}/(v_{\mathrm{A}} - v_{\mathrm{B}})$ (Chapter 4)
μ_{X}	fraction of X in the mobile phase of a chromatographic column
σ_{V_e}	standard deviation of the distribution of elution volume
τ_{ij}	intrinsic binding constant of the ith ligand to the jth acceptor
ϕ	fraction of cross-sectional area of chromatographic column occupied by mobile phase
ω	angular velocity in sedimentation
Ω	Kohlrausch regulating function

REFERENCES

ABRAMSON, H. A., MOYER, L. S., and GORIN, M. H. (1942) *Electrophoresis of proteins.* Reinhold, New York.

ACKERS, G. K. (1964) *Biochemistry, Easton* **3**, 723.

—— (1967) *J. biol. Chem.* **242**, 3026.

—— (1968) *J. biol. Chem.* **243**, 2056.

—— (1970) *Adv. Protein Chem.* **24**, 343.

—— and THOMPSON, T. E. (1965) *Proc. natn. Acad. Sci. U.S.A.* **53**, 342.

ADAMS, E. T., JR. (1967) Fractions No. 3, p. 1. Beckman (Spinco Division), Palo Alto.

—— (1969) *Ann. N.Y. Acad. Sci.* **164**, 226.

—— and FUJITA, H. (1963) *Ultracentrifugal analysis in theory and experiment* (ed. J. W. Williams), p. 119. Academic Press, New York.

—— PEKAR, A. H., SOUCEK, D. A., TANG, L. H., BARLOW, G., and ARMSTRONG, J. L. (1969) *Biopolymers* **7**, 5.

ADKINS, B. J. and YANG, J. T. (1968) *Biochemistry, Easton* **7**, 266.

ALBERTY, R. A. and MARVIN, H. H., JR. (1950) *J. phys. Colloid Chem.* **54**, 47.

ANDERSON, E. A. and ALBERTY, R. A. (1948) *J. phys. Colloid Chem.* **52**, 1345.

ANDREWS, P. (1966) *Br. med. Bull.* **22**, 109.

ATKINSON, D. E. (1965) *Science, N.Y.* **150**, 851.

BAGHURST, P. A., NICHOL, L. W., RICHARDS, R. J., and WINZOR, D. J. (1971) *Nature, Lond.* **234**, 299.

BALDWIN, R. L., DUNLOP, P. J., and GOSTING, L. J. (1955) *J. Am. chem. Soc.* **77**, 5235.

BARNETT, L. B. and BULL, H. B. (1960) *Archs Biochem. Biophys.* **89**, 167.

BELFORD, G. G. and BELFORD, R. L. (1962) *J. chem. Phys.* **37**, 1926.

BERAN, M. J. (1957) *J. chem. Phys.* **27**, 270.

BETHUNE, J. L. (1965*a*) *Biochemistry, Easton* **4**, 2691.

—— (1965*b*) *Biochemistry, Easton* **4**, 2698.

—— and GRILLO, P. J. (1967) *Biochemistry, Easton* **6**, 796.

—— and KEGELES, G. (1961*a*) *J. phys. Chem. Wash.* **65**, 433.

—— —— (1961*b*) *J. phys. Chem. Wash.* **65**, 1755.

—— —— (1961*c*) *J. phys. Chem. Wash.* **65**, 1761.

BLAKE, A. and PEACOCKE, A. R. (1968) *Biopolymers* **6**, 1225.

BLATT, W. F., ROBINSON, S. M., and BIXLER, H. J. (1968) *Analyt. Biochem.* **26**, 151.

BOOTH, F. (1950) *Proc. R. Soc.* **A203**, 514.

—— (1954) *J. chem. Phys.* **22**, 1956.

BRASWELL, E. (1968) *J. phys. Chem. Wash.* **72**, 2477.

BRUMBAUGH, E. E. and ACKERS, G. K. (1968) *J. biol. Chem.* **243**, 6315.

—— (1971) *Analyt. Biochem.* **41**, 543.

CANN, J. R. (1949) *J. Am. chem. Soc.* **71**, 907.

—— (1958) *J. Am. chem. Soc.* **80**, 4263.

—— (1959) *J. phys. Chem. Wash.* **63**, 210.

—— (1960) *J. biol. Chem.* **235**, 2810.

—— (1961) *J. Am. chem. Soc.* **83**, 4784.

—— (1966) *Biochemistry, Easton* **5**, 1108.

—— (1970) *Interacting macromolecules*, Academic Press, New York.

—— and BAILEY, H. R. (1961) *Archs Biochem. Biophys.* **93**, 576.

—— and GOAD, W. B. (1965*a*) *J. biol. Chem.* **240**, 148.
—— —— (1965*b*) *J. biol. Chem.* **240**, 1162.
—— KIRKWOOD, J. G., and BROWN, R. A. (1957) *Archs Biochem. Biophys.* **72**, 37.
—— and KLAPPER, J. A., JR. (1961) *J. biol. Chem.* **236**, 2446.
—— and PHELPS, R. A. (1957) *J. Am. chem. Soc.* **79**, 4672.
—— —— (1959) *J. Am. chem. Soc.* **81**, 4378.
CANNAN, R. K., KIBRICK, A., and PALMER, A. H. (1941) *Ann. N.Y. Acad. Sci.* **41**, 243.
CARR, C. W. (1953) *Archs Biochem. Biophys.* **46**, 417.
CHANGEUX, J-P., GERHART, J. C., and SCHACHMAN, H. K. (1968) *Biochemistry, Easton*, **7**, 531.
—— and RUBIN, M. M. (1968) *Biochemistry, Easton* **7**, 553.
—— THIÉRY, J., TUNG, Y., and KITTEL, C. (1967) *Proc. natn. Acad. Sci. U.S.A.* **57**, 335.
CHARLWOOD, P. A. (1950) *Biochem. J.* **46**, 312.
—— and ENS, A. (1957) *Can. J. Chem.* **35**, 99.
CHIANCONE, E. and GILBERT, G. A. (1965) *J. biol. Chem.* **240**, 3866.
—— GILBERT, L. M., GILBERT, G. A., and KELLETT, G. L. (1968) *J. biol. Chem.* **243**, 1212.
CLARKE, F. M., MASTERS, C. J., and WINZOR, D. J. (1970) *Biochem. J.* **118**, 325.
COLMAN, R. F. and FRIEDEN, C. (1966) *J. biol. Chem.* **241**, 3661.
COOPER, P. F. and WOOD, G. C. (1968) *J. Pharm. Pharmacol.* **20**, 150S.
COX, D. J. (1967) *Archs Biochem. Biophys.* **119**, 230.
—— (1969) *Archs Biochem. Biophys.* **129**, 106.
CRAIG, L. C. and KONIGSBERG, W. (1961) *J. phys. Chem. Wash.* **65**, 166.
CREETH, J. M. (1964) *Proc. R. Soc.* **A282**, 403.
—— and NICHOL, L. W. (1960) *Biochem. J.* **77**, 230.
—— and WINZOR, D. J. (1962) *Biochem. J.* **83**, 566.
DAVIES, M., NICHOL, L. W., and OGSTON, A. G. (1963) *Biochim. biophys. Acta* **75**, 436.
DEBYE, P. and HÜCKEL, E. (1923) *Phys. Z.* **24**, 185.
DECKER, R. V. and FOSTER, J. F. (1966) *Biochemistry, Easton* **5**, 1242.
DENBURG, J. and DE LUCA, M. (1968) *Biochem. biophys. Res. Commun.* **31**, 453.
DEVAULT, D. (1943) *J. Am. chem. Soc.* **65**, 532.
DISHON, M., WEISS, G. H., and YPHANTIS, D. A. (1966) *Biopolymers* **4**, 449.
—— —— —— (1967) *Biopolymers* **5**, 697.
DONNAN, F. G. (1923) *Chem. Rev.* **1**, 73.
DUNLOP, P. J. and GOSTING, L. J. (1955) *J. Am. chem. Soc.* **77**, 5238.
EDMOND, E., FARQUHAR, S., DUNSTONE, J. R., and OGSTON, A. G. (1968) *Biochem. J.* **108**, 755.
EISENBERG, H. and TOMKINS, G. M. (1968) *J. molec. Biol.* **31**, 37.
FLODIN, P. (1962) *Dextran gels and their applications in gel filtration.* Pharmacia, Uppsala.
FRIEDEN, C. (1967) *J. biol. Chem.* **242**, 4045.
—— and COLMAN, R. F. (1967) *J. biol. Chem.* **242**, 1705.
FUJITA, H. (1956) *J. chem. Phys.* **24**, 1084.
—— (1962) *The mathematical theory of sedimentation analysis.* Academic Press, New York.
GELOTTE, B. (1960) *J. Chromat.* **3**, 330.
GERHART, J. C. and PARDEE, A. B. (1962) *J. biol. Chem.* **237**, 891.
—— —— (1963) *Cold Spring Harb. Symp. quant. Biol.* **28**, 491.
—— and SCHACHMAN, H. K. (1968) *Biochemistry, Easton* **7**, 538.
GIDDINGS, J. C. and MALLIK, K. L. (1966) *Analyt. Chem.* **38**, 997.
GILBERT, G. A. (1955) *Discuss. Faraday Soc.* **20**, 68.
—— (1959) *Proc. R. Soc.* **A250**, 377.
—— (1960) *Nature, Lond.* **186**, 882.
—— (1963) *Proc. R. Soc.* **A276**, 354.
—— (1966*a*) *Nature, Lond.* **210**, 299.
—— (1966*b*) *Nature, Lond.* **212**, 296.
—— and JENKINS, R. C. L. (1959) *Proc. R. Soc.* **A253**, 420.

—— —— (1963) *Ultracentrifugal analysis in theory and experiment* (ed. J. W. Williams), p. 59. Academic Press, New York.
GILBERT, L. M. and GILBERT, G. A. (1961) *Nature, Lond.* **192**, 1181.
—— —— (1965) *Biochem. J.* **97**, 7C.
—— —— (1968) *The regulation of enzyme activity and allosteric interactions* (ed. E. Kvamme and A. Pihl), p. 73. Universitetsforlaget, Oslo.
—— and KELLETT, G. L. (1971) *J. biol. Chem.* **246**, 6079.
GLUECKAUF, E. (1955) *Trans. Faraday Soc.* **51**, 34.
GOLDBERG, R. J. (1953) *J. phys. Chem. Wash.* **57**, 194.
HALVORSON, H. R. and ACKERS, G. K. (1971) *J. Polym. Sci., A-2* **9**, 245.
HAMMES, G. G. and SCHIMMEL, P. R. (1965) *J. Am. chem. Soc.* **87**, 4665.
HARRIS, C. E., KOBES, R. D., TELLER, D. C., and RUTTER, W. J. (1969) *Biochemistry, Easton* **8**, 2442.
HENRY, D. C. (1931) *Proc. R. Soc.* **A133**, 106.
HERSH, R. and SCHACHMAN, H. K. (1955) *J. Am. chem. Soc.* **77**, 5228.
HIBBERD, G. E., OGSTON, A. G., and WINZOR, D. J. (1970) *J. Chromat.* **48**, 393.
HILL, A. V. (1910) *J. Physiol., Lond.* **40**, iv.
HOAGLAND, V. D., JR. and TELLER, D. C. (1969) *Biochemistry, Easton* **8**, 594.
HOCH, H. (1950) *Biochem. J.* **46**, 199.
HOUGHTON, G. (1963) *J. phys. Chem. Wash.* **67**, 84.
HOWLETT, G. J., JEFFREY, P. D., and NICHOL, L. W. (1970) *J. phys. Chem. Wash.* **74**, 3607.
—— (1972) *J. phys. Chem. Wash.* (in press).
HUMMEL, J. P. and DREYER, W. J. (1962) *Biochim. biophys. Acta* **63**, 530.
—— VER PLOEG, D. A. and NELSON, C. A. (1961) *J. biol. Chem.* **236**, 3168.
JEFFREY, P. D. and COATES, J. H. (1966) *Biochemistry, Easton* **5**, 489.
JOHNSTON, J. P. and OGSTON, A. G. (1946) *Trans. Faraday Soc.* **42**, 789.
JOSEPHS, R. and HARRINGTON, W. F. (1966) *Biochemistry, Easton* **5**, 3474.
—— —— (1967) *Proc. natn. Acad. Sci. U.S.A.* **58**, 1587.
—— —— (1968) *Biochemistry, Easton* **7**, 2834.
KEGELES, G. (1969) *Biopolymers* **7**, 83.
—— and JOHNSON, M. (1970) *Archs. Biochem. Biophys.* **141**, 59.
—— RHODES, L. and BETHUNE, J. L. (1967) *Proc. natn. Acad. Sci. U.S.A.* **58**, 45.
KELLER, R. A. and GIDDINGS, J. C. (1960) *J. Chromat.* **3**, 205.
KLOTZ, I. M. (1946) *Archs Biochem.* **9**, 109.
—— (1953) *The proteins* (eds. H. Neurath and K. E. Bailey), Vol. 1B, p. 727. Academic Press, New York.
—— WALKER, F. M. and PIVAN, R. B. (1946) *J. Am. chem. Soc.* **68**, 1486.
KOHLRAUSCH, F. (1897) *Annln Phys.* **62**, 209.
KOSHLAND, D. E., JR. (1958) *Proc. natn. Acad. Sci. U.S.A.* **44**, 98.
—— (1960) *Adv. Enzymol.* **22**, 45.
—— (1963) *Cold Spring. Harb. Symp. quant. Biol.* **28**, 473.
—— (1964) *Fed. Proc. Fedn. Am. Socs. Exp. Biol.* **23**, 719.
—— (1970) *The Enzymes* (ed. P. D. BOYER). Vol. 1, p. 342, Academic Press, New York.
—— and NEET, K. E. (1968) *A. Rev. Biochem.* **37**, 359.
—— NÉMETHY, G. and FILMER, D. (1966) *Biochemistry, Easton* **5**, 365.
KUBACKI, V., BROWN, K. D., and LASKOWSKI, M. (1949) *J. biol. Chem.* **180**, 73.
LANGERMAN, N. R. and KLOTZ, I. M. (1969) *Biochemistry, Easton* **8**, 4746.
LAURENT, T. C. and KILLANDER, J. (1964) *J. Chromat.* **14**, 317.
LEBOWITZ, J. and LASKOWSKI, M., JR. (1962) *Biochemistry, Easton* **1**, 1044.
LONGSWORTH, L. G. (1941) *Ann. N.Y. Acad. Sci.* **41**, 267.
—— (1943) *J. Am. chem. Soc.* **65**, 1755.
—— (1947) *J. phys. Colloid Chem.* **51**, 171.

—— (1959*a*) *Electrophoresis, theory, methods, and applications* (ed. M. Bier), p. 91. Academic Press, New York.
—— (1959*b*) *Electrophoresis, theory, methods, and applications* (ed. M. Bier), p. 137. Academic Press, New York.
—— and MACINNES, D. A. (1942) *J. gen. Physiol.* **25**, 507.
LUMRY, R. and RAJENDER, S. (1970) *Biopolymers* **9**, 1125.
MCKENZIE, H. A. and NICHOL, L. W. (1970) *Milk proteins.* (ed. H. A. McKenzie), Vol. 1, p. 267. Academic Press, New York.
MAKINO, R. C. and ROGERS, E. (1965*a*) *Archs Biochem. Biophys.* **109**, 499.
—— —— (1965*b*) *Archs Biochem. Biophys.* **109**, 560.
MARKUS, G., MCCLINTOCK, D. K., and BUSSELL, J. B. (1971) *J. biol. Chem.* **246**, 762.
MARTIN, W. G., COOK, W. H., and WINKLER, C. A. (1956) *Can. J. Chem.* **34**, 809.
MASSEY, V., HARRINGTON, W. F., and HARTLEY, B. S. (1955) *Discuss. Faraday Soc.* **20**, 24.
MASTERS, C. J., SHEEDY, R. J., WINZOR, D. J., and NICHOL, L. W. (1969) *Biochem. J.* **112**, 806.
MONOD, J., WYMAN, J., and CHANGEUX, J-P. (1965) *J. molec. Biol.* **12**, 88.
MOORE, C. H., HENDERSON, R. W., and NICHOL, L. W. (1968) *Biochemistry, Easton* **7**, 4075.
MYER, Y. P. and SCHELLMAN, J. A. (1962) *Biochim. biophys. Acta* **55**, 361.
NEURATH, H. and DREYER, W. J. (1955) *Discuss. Faraday Soc.* **20**, 32.
NICHOL, L. W. (1965) *Aust. J. Sci.* **27**, 342.
—— and BETHUNE, J. L. (1963) *Nature, Lond.* **198**, 880.
—— —— KEGELES, G. and HESS, E. L. (1964) *The proteins* (ed. H. Neurath), Vol. 2, p. 305. Academic Press, New York.
—— JACKSON, W. J. H. and SMITH, G. D. (1971) *Archs Biochem. Biophys.* **144**, 438.
——— and WINZOR, D. J. (1967*a*) *Biochemistry, Easton* **6**, 2449.
—— (1972a) *Biochemistry, Easton* **11** (in press).
—— O'DEA, K. and BAGHURST, P. A. (1972) *J. theoret. Biol.* (in press).
—— and OGSTON, A. G. (1965*a*) *Proc. R. Soc.* **B163**, 343.
—— —— (1965*b*) *J. phys. Chem. Wash.* **69**, 1754.
—— —— (1965*c*) *J. phys. Chem. Wash.* **69**, 4365.
—— —— (1967) *Proc. R. Soc.* **B167**, 164.
—— —— and RESCIGNO, A. (1970) *Proc. R. Soc.* **A317**, 153.
—— —— and WINZOR, D. J. (1967*b*) *J. phys. Chem. Wash.* **71**, 726.
—— —— —— (1967*c*) *Archs Biochem. Biophys.* **121**, 517.
—— and ROY, A. B. (1965) *Biochemistry, Easton* **4**, 386.
—— SAWYER, W. H. and WINZOR, D. J. (1969*a*) *Biochem. J.* **112**, 259.
—— SMITH, G. D. and OGSTON, A. G. (1969*b*) *Biochim. biophys. Acta* **184**, 1.
—— —— and WINZOR, D. J. (1969*c*) *Nature, Lond.* **222**, 174.
—— and WINZOR, D. J. (1964) *J. phys. Chem. Wash.* **68**, 2455.
—— —— (1965) *Biochim. biophys. Acta* **94**, 591.
OBERHAUSER, D. F., BETHUNE, J. L., and KEGELES, G. (1965) *Biochemistry, Easton* **4**, 1878.
OVERBEEK, J. T. G. (1950) *Adv. Colloid Sci.* **3**, 97.
PAYENS, T. A. J. and SCHMIDT, D. G. (1966) *Archs Biochem. Biophys.* **115**, 136.
PEDERSEN, K. O. (1958) *J. phys. Chem. Wash.* **62**, 1282.
PEPE, F. A. and SINGER, S. J. (1956) *J. Am. chem. Soc.* **78**, 4553.
—— —— (1959) *J. Am. chem. Soc.* **81**, 3878.
PHELPS, R. A. and CANN, J. R. (1956) *J. Am. chem. Soc.* **78**, 3539.
PORATH, J. (1963) *Pure appl. Chem.* **6**, 233.
RAFTERY, M. A., DAHLQUIST, F. W., CHAN, S. I., and PARSONS, S. M. (1968) *J. biol. Chem.* **243**, 4175.
RAO, M. S. N. and KEGELES, G. (1958) *J. Am. chem. Soc.* **80**, 5724.
REITHEL, F. J. (1963) *Adv. Protein Chem.* **18**, 123.
RICHARDS, E. G. and SCHACHMAN, H. K. (1957) *J. Am. chem. Soc.* **79**, 5324.

ROARK, D. E. and YPHANTIS, D. A. (1969) *Ann. N.Y. Acad. Sci.* **164**, 245.
RONALDS, J. A. and WINZOR, D. J. (1969) *Archs Biochem. Biophys.* **129**, 94.
SCATCHARD, G. (1949) *Ann. N.Y. Acad. Sci.* **51**, 660.
SCHACHMAN, H. K. (1959) *Ultracentrifugation in biochemistry*. Academic Press, New York.
SCHOLTEN, P. C. (1961) *Archs Biochem. Biophys.* **93**, 568.
SCHUMAKER, V. N. and ROSENBLOOM, J. (1965) *Biochemistry, Easton* **4**, 1005.
SCHWERT, G. W. (1949) *J. biol. Chem.* **179**, 655.
SIEGEL, L. M. and MONTY, K. J. (1966) *Biochim. biophys. Acta* **112**, 346.
SINGER, S. J., PEPE, F. A., and ILTEN, D. (1959) *J. Am. chem. Soc.* **81**, 3887.
SMITH, E. L., KIMMEL, J. R., and BROWN, D. M. (1954) *J. biol. Chem.* **207**, 533.
SMITH, G. D. and NICHOL, L. W. (1971) *Aust. J. biol. Sci.* **24**, 955.
SOPHIANOPOULOS, A. J. and VAN HOLDE, K. E. (1964) *J. biol. Chem.* **239**, 2516.
SQUIRE, P. G. (1964) *Archs Biochem. Biophys.* **107**, 471.
—— MOSER, P. and O'KONSKI, C. T. (1968) *Biochemistry, Easton* **7**, 4261.
STADTMAN, E. R. (1966) *Adv. Enzymol.* **28**, 41.
STEINBERG, I. Z. and SCHACHMAN, H. K. (1966) *Biochemistry, Easton* **5**, 3728.
STEINER, R. F. (1952) *Archs Biochem. Biophys.* **39**, 333.
SVEDBERG, T. and PEDERSEN, K. O. (1940) *The ultracentrifuge*. Oxford University Press, London and New York.
SVENSSON, H. (1943) *Ark. Kemi Miner. Geol.* **17A**, No. 14, 1.
TEN EYCK, L. F. and KAUZMANN, W. (1967) *Proc. natn. Acad. Sci. U.S.A.* **58**, 888.
TIMASHEFF, S. N. and KIRKWOOD, J. G. (1953) *J. Am. chem. Soc.* **75**, 3124.
TISELIUS, A. (1930) *Nova Acta R. Soc. Scient. upsal.* [4] **7**, No. 4, p. 1.
—— (1932) *Kolloid. Z.* **59**, 306.
—— and SVENSSON, H. (1940) *Trans. Faraday Soc.* **36**, 16.
TOMBS, M. P. (1957) *Biochem. J.* **67**, 517.
TOWNEND, R., WINTERBOTTOM, R. J., and TIMASHEFF, S. N. (1960) *J. Am. chem. Soc.* **82**, 3161.
TRAUTMAN, R. and SCHUMAKER, V. N. (1954) *J. chem. Phys.* **22**, 551.
UMBARGER, H. E. (1964) *Science, N.Y.* **145**, 674.
VAN HOLDE, K. E. (1962) *J. chem. Phys.* **37**, 1922.
VELICK, S. F., PARKER, C. W., and EISEN, H. N. (1960) *Proc. natn. Acad. Sci. U.S.A.* **46**, 1470.
WALDMANN-MEYER, H. (1960) *J. biol. Chem.* **235**, 3337.
WEBER, G. (1965) *Molecular biophysics* (eds. B. Pullman and M. Weissbluth), p. 369. Academic Press, New York.
—— and ANDERSON, S. R. (1965) *Biochemistry, Easton* **4**, 1942.
WEBER, K. (1968) *Nature, Lond.* **218**, 1116.
WILEY, D. C. and LIPSCOMB, W. N. (1968) *Nature, Lond.* **218**, 1119.
WINZOR, D. J. (1966) *Biochem. J.* **101**, 30C.
—— (1970) *Biochim. biophys. Acta* **200**, 423.
—— LOKE, J. P. and NICHOL, L. W. (1967) *J. phys. Chem. Wash.* **71**, 4492.
—— and NICHOL, L. W. (1965) *Biochim. biophys. Acta* **104**, 1.
—— and SCHERAGA, H. A. (1963) *Biochemistry, Easton* **2**, 1263.
—— —— (1964) *J. phys. Chem. Wash.* **68**, 338.
WYMAN, J. (1964) *Adv. Protein Chem.* **19**, 223.
—— (1967) *J. Am. chem. Soc.* **89**, 2202.

AUTHOR INDEX

Page numbers in *italic* indicate a literature reference

SUBJECT INDEX